THE
UFOs
THAT NEVER
WERE

THE
UFOs
THAT NEVER
WERE

JENNY RANDLES

ANDY ROBERTS

DAVID CLARKE

LONDON
HOUSE

First published in Great Britain in 2000 by
LONDON HOUSE
114 New Cavendish Street
London W1M 7FD

A catalogue record for this book is available
from the British Library

ISBN 1 902809 35 1

Jacket design by Neil Straker Creative.

Edited and designed by DAG Publications Ltd, London.
Printed and bound by
Creative Print and Design (Wales), Ebbw Vale.

CONTENTS

CONTENTS
of boxed features

ACKNOWLEDGEMENTS

Dr David Clarke wishes to thank

Nick Pope, for allowing me to use quotes from private correspondence; Nick Redfern for obtaining the MoD file on the Alex Birch photos; Ian Lindsey and Mark Clarke of HM Coastguard; Mike France and Phil Shaw of the Peak District Mountain Rescue Organisation; Flight Lt. Tom Rounds and Alan Patterson of the RAF Press Office, MoD; Professor Mark Bailey of Armagh Observatory; Martin Jeffrey, for help and access to files on the Howden Moor case; Jen Topping of the *Stornway Gazette*; Glenn Ford of the British Geological Survey, Edinburgh; Gloria Dixon, Tim Matthews and Mike Wootten for material and encouragement; David Brownlow; Stuart Dixon; Alex Birch for allowing use of his UFO photograph and material; Jerome Clark and Eddie Bullard for vital airship references; Helen Jackson MP for asking the right questions in parliament; Calum MacDonald MP for aid with the Isle of Lewis case; South Yorkshire Police for valued help and co-operation in the investigation of the Howden Moor incident.

Jenny Randles wishes to thank

Peter Day, for his years of quiet co-operation; Paula, Nina and Tracey for their patience; Peter Southerst and all the staff at Kodak UK for assistance above and beyond; Peter Warrington for posing the right questions; Roger Stanway, Ken Phillips and others at BUFORA who tried hard to sort out this case; Steuart Campbell, who helped, Marty Moffatt and SCUFORI, whose painstaking work was an inspiration; Brenda Butler and Dot Street, without whom the world would not have heard of Rendlesham Forest; Ian Ridpath and James Easton, without whom the world might not have found out what really happened in there; Ray Boeche, Scott Colborne and Barry Greenwood for their commitment and co-operation; Sally Rayl for her excellent interviews; all the witnesses, especially Charles Halt, John Burroughs, Mal Scurrah and Jim Penniston for their data; Philip Mantle and David Alpin for helping me find some important bits of it; Ralph Noyes for pointing us subtly in the right direction.

Andy Roberts wishes to thank

My co-authors, Dave and Jenny, for having the courage to speak out against the madness which grips UFOlogy; Robert Moore and Dave Kelly for use of their archival materials; Graham Birdsall for permission to quote widely from YUFOS case reports; Jerry Garcia and Robert Hunter for the inspiration, and Christine for being herself.

INTRODUCTION

Why on earth write a book about UFO cases that have been solved? Surely all interest in the mystery stems from its unexplained nature and because of that elusive prospect that aliens from a distant galaxy might be coming to our world. To take away the mystique and dare to find mundane solutions for a case is rather like being the biggest party pooper of all time. It might also be considered a very good way to kill off the golden goose since UFOs have provided a lucrative living for many writers, journalists, TV networks and researchers who can earn large sums and give back much excitement to the millions.

Of course, the fact that a subject is popular is not an excuse for promoting things that can be reasonably shown to be untrue. There is an obligation to take incredible claims and subject them to overly strict levels of investigation – being only forced kicking and screaming into considering that they might be paranormal. Too often UFOs are regarded as supernatural merely on the pretext that this provides the better story. And much of our society still thrives on good story telling if not now whilst sitting around the camp fire, then through the far more insidious means of TV chat shows and tabloid revelations.

What needs to be stressed immediately is that this is not a book that attempts to disprove the existence of UFOs. All three of its authors are long-term, dedicated UFOlogists who on some level *do* believe there are unsolved cases that may offer new knowledge. In that sense, we believe in UFOs and are not pretending that all cases can be wished away. We do not agree on every aspect of the subject, although we are all very sceptical of alien theories. But we are positively not debunkers out to shoot down every case in sight. If an incident appears unexplained – and some of them in our time most certainly do – we are more than happy to stand up and say so.

Nevertheless, as UFO researchers with a collective experience of some 75 years (possibly the most ever marshalled for the authors of such a book!), we have recognised some rules of the game along the way. The most fundamental of all is this: the majority of UFO sightings that you read about or watch on dramatic TV reconstructions are not UFO encounters at all. They are what we call IFOs – Identified Flying Objects.

IFOs have been reported as being UFOs as a result of a host of factors, including misperception, human psychology, social pressures, media hype, investigator incompetence and sometimes a combination of circumstances that triggers a seemingly incredible encounter.

Few serious UFOlogists dispute the figure that around 95 per cent of all reported UFO sightings turn out to be IFOs. Since in Britain some 300 cases per year are recorded and world-wide the total is several thousand, the sheer number of IFO cases during the long history of the UFO movement is over-whelming. Yet this huge data base is frequently ignored. The media work on the premise that only unsolved cases are worth publishing. Indeed, very often the hallowed status of a case as a great modern mystery owes more to its promotion and to the overly excitable wing of the UFO community than it does to witnesses or the facts. This can mean that seeking resolution leads one on a pilgrimage through lands filled with sacred cows, sniping away at the words and deeds of one's colleagues in order to get to the truth.

UFOlogists, too, often consider it a failure to have found an explanation because mentally they are geared-up to seek out proof of aliens. Witnesses only rarely want an answer, because they may wrongly assume that to find one infers either dishonesty or stupidity on their part. It frequently presumes neither. The best IFO cases would fool anyone in the world, regardless of status. Airline pilots, police officers, politicians and science professors have all been victims. But to work back towards an answer requires skill and dedication. It can be, as writer Peter Brookesmith terms it, "dark and lonely work." It is certainly a task that brings few friends or offers of support. However, integrity and good practice demand it to be done. The starting point must be witness testimony, sought out immediately – before the elements of contamination get in the way. There are all too many of these, from human preconceptions to UFOlogists who, incredibly at times, show Identikit photos or drawings of UFOs and aliens in an effort to get a witness to recall what he or she saw. This tactic, sadly, leads more to misperception than to truth.

We have attempted to base this book, as far as possible, on our own first-hand investigations. As a result, each of the main chapters feature a case that one or more of the authors spent much time researching, often for years, or even decades. The cases span the second half of the 20th century (and even go back into the 19th in one example). Although including incidents from the 1990s, it is a consequence of what we do that many date back to the 1970s or 1980s. Good UFO research does not happen overnight. It can often take much time to hack your way through the undergrowth towards an acceptable truth.

UFOlogy is also a dynamic, interactive field. We have certainly had the fortune to work with, and to be given access to, research carried out by a number of other dedicated investigators, some of them UFOlogists, but many eschewing that title. Wherever possible we have acknowledged that debt.

It is inevitable that by basing this book upon our own work the outcome biases the content towards British cases, and those from Northern England, where all three authors reside. However, we were careful not to limit ourselves in that way. As such, half the main chapters feature cases from beyond this remit, notably Scotland, Wales and Southern England.

In order to add an international dimension we also include a good selection of material from the USA. Our most in-depth case study in this book involves American witnesses and facilities, even though it occurred in the UK. We have also included a series of short reports, which serve to analyse some of the most respected cases in UFO history. This adopts a similar format to that attempted in greater detail within the main chapters.

These boxes feature cases from several countries, including a large number from the USA. We hope this demonstrates that our quest is not parochial and of relevance only to the UK. The lessons of the British sightings are equally valid beyond these shores. Overseas readers might derive even more fascination from 'playing detective' whilst reading them because they may be less familiar.

This book is not written with any great pretensions in mind. It is not aimed to demolish UFOs, or to disprove that something strange ever happens. It is not a diatribe against sloppy UFO investigation or hype practised by the media. Any criticism included is constructive, not destructive. We do not single out individuals or groups for criticism or infer that they are universally incompetent, but simply try to piece evidence together as a way to find truth. We are all seeking that truth in our own special way.

Another unrecognised factor is this. IFO cases are sometimes a lot more interesting than the stories left in limbo with that rather meaningless tag word "unidentified." To be unidentified is not to be insoluble. A UFO is an "unidentified" flying object and not, as is all too widely assumed, an *unidentifiable* one. In fact, the cases that remain unexplained may do so only temporarily until someone finds the key to unlock the answer. In this book we present some solved cases that were regarded as impressive and unidentified for decades. Indeed, in some quarters they are still viewed that way today.

The failures – if that word can ever be fairly applied to an amorphous and contentious subject such as UFOs – are the cases left unresolved. For they are not proof of UFO reality. Potentially, all are open to explanation at some point in the future. IFO cases are properly the success stories because they have a beginning, a middle and an end, which is true of all too little in the myth-inspired, guesswork-riddled UFO scene. They reflect what is good about UFOlogy – the hard work, the determination and the quest to find an answer come what may. Often this takes fortune, and always perseverance. But there are some good UFOlogists who are practising these skills.

Sadly, these people often go unsung because the media prefer to publish daft quotes from the more extreme believers and to treat UFOlogy as amusing filler and not a real story. Many large unwieldy groups fulfil the function of a social club for enthusiasts, but are rarely geared-up to tackle the hard issues of UFOlogy. It is easy to see why. To do so would often place them in conflict with *The X Files*-inspired aspirations of the members who are the lifeblood of a group. They want aliens and conspiracies, whereas the facts often dictate the reverse.

As such – and as computer scientist and UFOlogist Jacques Vallee rightly puts it – "no major advances have been made by groups, but instead by individuals and small teams." They have no magazines and conferences to fund, and no elections to face where their fate is decreed by how well they champion the fight to prove that ET has landed. Indeed, the less rigorous end of the UFO movement actually can see undue effort to solve cases as proof that we are government agents out to mislead the public by emphasising all the wrong aspects. What we see as information – necessary for the public to view to make a balanced judgement on these cases – they may regard as disinformation.

This argument is weird but understandable. At times, the powers-that-be have played fast and loose with the UFO evidence, and their own actions have bred paranoia and probably unjustified cries of "Cover-up!" The authorities need to look again at the needless secrecy imposed on data they claim is of no major significance. We often support that diagnosis, but by failing to make this data public, governments can shoot down their own rational arguments.

What can we do to reassure the reader that we are not some puppet of the intelligence community? Nothing, frankly. If we say we are not, those who already know that (hopefully most readers!) would not need telling. But others who do not believe this will hardly take our word for it anyway.

In the crazy world of UFO buffs you either support the "great truth" – that a massive alien invasion is underway – or are considered with deep suspicion even for being interested in this field. Yet it is not only possible to be fascinated by UFOs whilst not basing all your work on the belief that the aliens are here, but in our view it is essential. Beliefs get in the way of common sense and objectivity, qualities in far too short a supply in the world of UFO study. The only way you can properly study the UFO evidence is if you have no vested interest, one way or the other, in any specific outcome. One thing alone is absolute. Truth.

So we can do nothing except allow our work to speak for itself. You can judge whether we are sincere and report on what we find, as is certainly our intention. However, it is important to emphasise one other point. This book contains our research, our conclusions and our convictions about the outcome of a case. As you will see, the term "solved" can be a relative one. In certain cases, we doubt any reasonable-minded person will disagree with our verdict. The evidence appears overwhelming. At other times, even the three authors are not wholly of one mind about the status of an incident, although we agreed to include only examples where we feel the balance of probability is that no "real UFO" was involved.

There will also certainly be cases where other UFOlogists, perhaps even witnesses, dispute our findings. That is fair enough. We do not live in a perfect world where all the answers are cut and dried. In UFOlogy, you have to balance pros and cons, and weigh probabilities when making a judgement call about a case. This is what we do here, but we are happy to debate the evidence in a friendly, civilised fashion with anyone and fully accept the premise that we can be wrong. The only request we make in return is that our critics admit to the same human failing.

You can, if you wish, play a little game with yourself as you read this book. We have written up the cases almost like a detective thriller. First, you will discover the story of the case as it was publicly revealed, not uncommonly via media headlines or other vociferous forces in the vigilante UFO world. Then we tread the path of investigation that led towards a probable solution and you can sift the clues, make notes about the various theories and become a DIY Miss Marple or Sherlock Holmes as you try to beat the authors to the answer. In that sense, perhaps we have written the first ever UFO whodunnit, except that here, of course, we have a whatdunnit? Given that the culprits can vary from a twinkling star to a grey-skinned entity from Alpha Centauri, there is no shortage of suspects for you to evaluate.

In this book, we have decided to put right decades of disinterest in evidence that says a great deal about the UFO world. In these pages, any investigator worthy of that name will find lessons to apply to his or her work. Anyone truly interested in what lies behind the UFO mystery needs to give careful thought to the implications of these tales. And anyone not excited by the prospect of pitting your own wits against a strange case and seeking to unravel the causes that led up to it probably should not pursue an interest in UFOlogy.

If nothing else, UFO research is a struggle to find meaning within the incredible and a battle to secure evidence that can be widely agreed upon. When hunting little green men, that is a thankless task since the prospect of taking one alive (or dead!) seems to diminish every day, but with cases such as these it can be much more rewarding. For here there is a definite prospect of succeeding in one's goal.

To be a UFOlogist you need to be part social worker, a detective, a lateral thinker, a logic puzzle addict and have dogged commitment to wade through a myriad possibilities towards the truth. We suspect that you will need many of these qualities to best enjoy this book. Good hunting!

Dr David Clarke, Jenny Randles, Andy Roberts. May 1999

1

ONCE UPON A TIME
IN THE WEST

DAVID CLARKE

"The great airship craze of the 1890s affords students of folklore and myth an unparalleled opportunity to study, in Richard Wilbur's phrase, 'at what cross-purposes the world is dreamt.' With the possible exception of abduction, it contains in embryonic form every theme present in our own century's encounter with UFOs." – Jeff Gorvetzian, *Fortean Times*.[1]

Writers of UFO history tend to begin their storybooks in 1946 and 1947 when the first wave of 'flying saucer' sightings swept across the USA and the world. Great significance has been read into the appearance of the saucers at the time of the first atomic tests and the development of new jet aircraft and rockets which placed mankind on the launch pad into space. This 'official' version of the subject is itself socially constructed and conceals the fact that the foundations of the UFO myth were laid many years before Kenneth Arnold's fateful sighting in 1947. They can be traced to the images of alien invaders in comic books, to the *War of the Worlds* radio broadcast which caused panic in the streets of New Jersey in 1938, and even further back to the novels of Jules Verne and great airship scares which plagued Europe and North America at the turn of the century.

The great airship craze of 1896–97 was certainly the precursor of all recent UFO waves in that it featured prototypes in narrative form of every single aspect of the 'modern' phenomenon from 'close encounters,' to animal mutilations, crash-landings of strange aircraft and even versions of the recent claims of contactees, abductees and conspiracy theorists. The airship sightings have to be seen in the context of the overwhelming belief at that time in the USA that an inventor had secretly developed the first practical airship which he was in the process of testing. In the years building up to the wave, popular science fiction and literature were replete with 'secret inventor' stories and claims that the first airship or heavier-than-air flying machine would soon be perfected.[2]

The last twenty years of the 19th century saw rapid advances in technology which placed the dream of aerial navigation within the grasp of mankind for the first time. It was a desire also kindled in fiction such as Jules Verne's *Robur the Conquerer* published in 1886, and its sequel *Master of the World* (1905). At the same time, astronomers speculated about Martian canals whilst H. G. Wells visualised an alien invasion of earth in his *War of the Worlds*, first published in 1898. Meanwhile, numerous backwoods inventors in America were known to be hard at work on their own flying machines. Several had obtained patents for craft by the late 1890s. In the event, it was 1903 before the Wright Brothers took the first faltering steps into the air with their flimsy aeroplane at Kitty Hawk, North Carolina. Earlier attempts at heavier-than-air flight were crude and usually ended in disaster.

Aviation historian Charles Gibbs-Smith has stated unequivocally that the only airborne vehicles carrying passengers which could possibly have been seen over North America in 1896–97 were free-flying balloons. He said, "No form of dirigible (i.e. a gasbag propelled by an airscrew) or heavier-than-air flying machine was flying – or indeed could fly – at this time."[3] Any survey of a random sample of the US newspapers from November–December 1896 and February to May 1897 would give readers a completely different impression.

For four months that winter, the columns of American newspapers were filled daily with the most fantastic and bizarre reports describing the movements of a 'mystery airship.' Folklore historian Eddie Bullard has said a conservative estimate of the number of individual sightings during the wave would be in excess of 100,000, with the vast majority of these taking place at night.[4] Observers described the 'airship' as cigar-shaped, 60 feet in length, silver in colour and equipped with a variety of wings, sails and propellers. One account from Beaumont, Texas, at the height of the hysteria described the ship as "... shaped like a Mexican cigar, large in the middle, and small at both ends, with great wings, resembling those of an enormous butterfly. It was brilliantly illuminated by the rays of two great searchlights, and was sailing in a south-easterly direction, with the velocity of the wind, presenting a magnificent appearance."[5]

Many of the sightings which took place at night closely tally with modern UFOs in terms of the descriptions of coloured lights surrounding dark objects in the sky, swift and erratic movements, and blinding searchlights directed towards the earth. But lacking are descriptions of the flying disk or saucer-shaped craft found in modern waves as observers in the 1890s saw what they expected to see in the sky, namely a prototype form of lighter-than-air dirigible. Consequently, the reports describe large cigar-shaped objects often with whirring propellers, great fans or even wings, which rose and fell as the airship progressed through the sky.

The sightings began in California in November, 1896, and spread eastwards across the Midwest, ending in May the following year with a brief pause during the Christmas period.

They appear to have been triggered by a series of speculative articles printed in US newspapers just before the craze for seeing airships began. In one, a reporter in the *Detroit Free Press* claimed that a New York inventor was about to test "an aerial torpedo boat." Sixteen days later, the California paper *Sacramento Bee* published a telegram from a New York man which claimed he and two friends were about to board an airship which would be flown across the Continent to California, which he promised to reach in two days.[6] Coincidentally, that very night, many witnesses in the same city reported seeing "an electric arc lamp propelled by some mysterious force" floating on the horizon.

As hundreds of people gathered on street-corners, some reported seeing a cigar-shaped craft which appeared to rise in the sky and then drop towards rooftops as it flew across Sacramento. Some people even claimed to hear voices from the sky shouting: "Lift her up quick! You are making directly for that steeple." Another man maintained that the airship was "propelled by fan-like wheels operated by four men, who worked as if on bicycles."[7]

Despite the circumstantial nature of stories such as this one, sensational newspaper hoaxes were a major feature of the craze. In these cases, writers and correspondents usually gave the game away at the end of the account by saying they were writing from an asylum for the insane, or something of that kind. Newspaper proprietor William Randolph Hearst attacked this form of 'yellow journalism' in an editorial which said: "Fake journalism has a lot to answer for, but we do not recall a more discernible exploit … than the persistent attempt to make the public believe that the air in this vicinity is populated with airships. It has been manifest for weeks that the whole airship story is pure myth."[8]

Typical of the stories printed by the US newspapers in 1897 was that of two Illinois farmhands who came upon a landed craft. They were informed by its pilot, a bearded scientist, that it was a new invention which had been flown only at night in order not to attract attention. The aeronaut told them he had left the town of Quincy, 100 miles to the west, only half an hour earlier, an impossible speed for any known aerial craft of that era.[9] Research by UFOlogists nearly a century later found that the two men named in the account never existed, but the story was typical of the yarns which were printed in 1897 describing meetings with the 'aeronauts.'[10]

Like modern contact claims, the majority of the encounters with aeronauts involved a single witness and were said to have taken place at night in isolated areas, when the pilot or pilots were found at work busily repairing their landed craft. The air of mystery and secrecy which surrounded the world of early aviation, and the popular belief that inventors were secretly

THE KENNETH ARNOLD SIGHTING

When Kenneth Arnold taxied onto the runway at the Chehalis, Washington, airfield on June 24, 1947, he had no idea how momentous his flight was to become. Whilst searching for a downed transport plane, his attention was drawn by two bright flashes. These were caused by nine objects flying in formation, weaving in and out of the Cascade Mountains, sun glinting on their metallic surfaces.

Arnold described the objects as crescents, with wings but no tails. They were like nothing he had ever seen before. His sighting lasted one minute, forty-two seconds, during which time he used mountain peaks as marker points to determine their speed as being a fantastic 1,700 mph. Upon landing at Pendleton, Oregon, Arnold described their motion to a newsman as follows: 'They flew like a saucer would if you skipped it across water'. An imaginative sub-editor coined the term "flying saucer" and the modern era of UFOs was born.

Is it real?
● There is no reason to doubt that Arnold saw 'something.'
● Following newspaper reports of Arnold's sighting, at least twenty other UFO sightings were reported from the same date.
● One report came from a miner, who described a number of round, metallic-looking discs. This occurred in the afternoon … before Arnold's sighting.

Is it solved?
● Arnold did not see the classic domed flying saucer-type of UFO. But the media-generated term "flying saucer" created a frame of reference for many subsequent sightings.
● No evidence has arisen for a positive identification of Arnold's objects. The belief that they were alien craft is just one of many explanations and

not widely supported even by UFOlogists. Secret military jets, missiles, meteors and birds have all been suggested. Arnold at first believed they were jets, but later concluded that UFOs were "space animals" living in the upper atmosphere.

Conclusion
No-one knows what Kenneth Arnold saw, but were it not for his historic sighting, it is certain the term "flying saucer" would never have existed. His experience may have been the genesis of the flying saucer age, but its origin is unknown and its meaning mutable. Whilst it is not a significant case offering proof of real UFOs, countless witnesses and UFOlogists have projected their hopes, dreams and fears by way of the image a journalist built around it.

Further reading
"What Happened On June 24, 1947." pp. 28–34. Arnold, K. *UFO 1947–97*. Ed. Evans, H. and Stacy, D. John Brown Publishing, 1997.

perfecting their craft, helped to foster public acceptance of the reality of the airship. None of the reports was independently investigated, but coming at a time when stories about life on Mars were rife, there were some who suggested the airship could have been a visitor from another planet. In his account of the wave, the indefatigable collector Charles Fort said it was clear that practical jokers, bright stars and planets played a large role in triggering the sightings, but added that "... against such an alliance as this, between the jokers and the astronomers, I see a small chance for our data. The chance is in the future. If, in April 1897, extra-mundane voyagers did visit this earth, likely enough they will visit again."[11]

UFOlogists who in the 1960s and 1970s re-discovered and carefully re-constructed the 1896–7 wave from dusty newspaper archives were impressed by the apparent similarities of the descriptions of the craft to those observed during the 'flying saucer' era. At the same time, they were baffled by the claims of those who said they had met the airships' pilots (the 'aeronauts'). These stories were fantastic, even in comparison with late 20th century UFO contacts. Not only were many of the airship crews described as ordinary American citizens, but there were also descriptions of oriental-type dwarfs, giant hairy humanoids and eccentric-looking gentlemen, stories which clearly owed more to the imaginations of Victorian story-tellers than to reality.

Martian UFO occupants in 1897

One yarn in particular, printed by a newspaper in California in November 1896, made what could be the first link between aerial craft and aliens, 50 years before Kenneth Arnold's sighting. It also emphasises the fact that a popular image of alien beings visiting earth in strange flying machines and attempting to kidnap humans existed in popular fiction many years before the 'flying saucer' age.

The story, told by Col H. G. Shaw, a former member of the staff of a California newspaper, described how he and a companion were riding near the town of Lodi when they encountered three strange beings, a scene which caused their horses to snort in terror. The creatures were more or less human-like in appearance, but were seven feet in height and very slender with small, delicate, nail-less hands and long, narrow feet. They appeared to be friendly, "without any sort of clothing, but were covered with a natural growth ... as soft as silk to the touch, and their skin was like velvet. Their faces and heads were without hair, the ears were very small ... while the eyes were large and lustrous."[12]

Attempts to communicate with the beings, who carried a luminous egg-shaped light, came to nothing, Suddenly, they attempted to lift Col Shaw, "probably with the intention of carrying me away," but lacked the strength to move him or his friend. When he touched one of the creatures under the elbow, Col Shaw noticed it was lighter than one ounce in weight. At this point, one of the beings flashed their light towards a bridge where a 150-feet long airship was waiting.

Col Shaw wrote: "The three walked rapidly towards the ship with a swaying motion, their feet only touching the ground at intervals of about fifteen feet, and with a little spring they rose to the machine, opened a door in the side, and disappeared within." His tongue planted firmly in his cheek, Col Shaw concluded that the creatures they saw "were inhabitants of Mars, who had been sent to earth for the purpose of securing one of its inhabitants." Col Shaw went on to describe the claims of earthly secret inventors, who had claimed credit for the airship sightings, as "clumsy fakes."[13]

This story, like many others, was clearly a newspaper hoax, but displayed all the elements of the modern UFO phenomenon in the melting pot of popular fiction and imagination 50 years before the flying saucer craze began.

Today we know there was no real airship or heavier-than-air flying machine in existence at the latter end of the 19th century in North America. There had been a number of ineffective attempts to launch dirigibles (steerable balloons) in Europe as early as 1852, but these had not progressed significantly until the first successful flight of a Zeppelin airship over Lake Constance in 1900. However, these early and faltering steps were accompanied by voluminous fictional accounts describing advanced aerial craft and images of future wars in the air. In addition, newspapers and magazines from this period were full of drawings and cartoons depicting precisely the sort of aerial flying machines that were reported during the airship flaps. This led sociologist Robert Bartholemew to conclude, "The 1896–97 airship wave is viewed as a case of collective wish-fulfilment as a response to rapid sociotechnological strains and to rumours that someone had invented the world's first practical airship."[14]

During the 1897 craze, newspapers in almost every Midwest state carried stories chronicling the movements of the airship. The wave of hysteria which surrounded the sightings appears to have spread from newspaper to newspaper via the railroad telegraph stations, which give the impression that the craft was moving slowly eastwards across the United States. Indeed, in the 1960s, a number of retired railroad telegraph-operators came forward to state that the

1897 sightings had been part of a huge hoax invented by bored employees in the Midwest. They spread fantastic accounts of the airship's movements from one state to the next by means of the telegraph system. Other reports were the result of newspaper hoaxes, some of them so well planned and convincing that they were still fooling UFOlogists 100 years later. In an age when the public was less jaded than today, many of these fantastic yarns were enthusiastically adopted and retold.

The calfnapping airship hoax

Possibly the most successful hoax of that period came from the tiny settlement of Yates Center in Kansas. It was here that a rancher, Alexander Hamilton, signed an affidavit to the effect that on the night of April 19, 1897, he saw the airship crew steal a cow from his yard. Hamilton had impressive credentials, at one time being a member of the House of Representatives for his state, but he was also a keen practical joker. He even persuaded ten of his friends, including the local sheriff and his deputy, to sign an affidavit attesting to his honesty. Hamilton's original account of the sighting appeared in the Yates Center *Farmer's Advocate* on April 23, 1897.

The story was reproduced in a number of larger papers, including the *Kansas City Times* on April 27, 1897, and led to a flood of inquiries from newspapers across the country. The *Times'* account said that many people laughed at the stories about the airship, "but the thing is no joke to farmer Alexander Hamilton, who resides near Yates Center, Woodson county." It reported that the airship not only appeared in plain view of Hamilton and his family, frightening them out of their wits, but added that the captain of the 'vessel' had the nerve to swoop down upon the cow-lot and steal a three-year-old heifer.

The *Times'* writer added: "At any rate, that is what Hamilton says, and the Yates Center *Advocate*, which publishes the story, vouches for his honesty and great love of truth. In addition, a dozen well-known citizens, including State Oil Inspector E. V. Wharton, Sheriff M. E. Hunt and Banker H. H. Winter, testify that Hamilton's reputation for truth and veracity has never been questioned." Hamilton's account followed:

Last Monday night about half past 10 o'clock, we were awakened by a noise among the cattle. I rose, thinking perhaps my bull dog was performing some of his pranks, but upon going to my cow lot, to my

21

utter astonishment, an airship was slowly descending over my cow lot about forty rods from the house. Calling Old Heslip, my tenant, and my son Wall, we seized some axes and ran to the corral. Meantime, the ship had been gently descending until it was not more than thirty feet above the ground, and we came up to within fifty yards of it.

It consisted of a great cigar-shaped portion, possibly 300 feet long, with a carriage underneath. The carriage was made of panels of glass or other transparent substance, alternating with a narrow strip of some material. It was brilliantly lighted within and everything was clearly visible. There were three lights, one light an immense searchlight and two smaller, one red and the other green. The large one was susceptible of being turned in any direction.

It was occupied by six of the strangest beings I ever saw. There were two men, a woman, and three children. They were jabbering together, but we could not understand a syllable they said. Every part of the vessel which was not transparent was of a dark reddish color. We stood mute in wonder and fright, when some noise attracted their attention and they turned their light directly upon us.

Immediately upon catching sight of us, they turned on some unknown power, and a great turbine wheel about thirty feet in diameter, which was slowly revolving below the craft, began to buzz, sounding precisely like the cylinder of a separator, and the vessel rose as lightly as a bird. When about 300 feet above us, it seemed to pause and hover directly over a three-year-old heifer, which was bawling and jumping, apparently fast in the fence. Going to her, we found a cable about half an inch in thickness, made of the same red material, fastened in a slip knot around her neck, one end passing up to the vessel and tangled in the wire. We tried to get it off, but could not, so we cut the wire loose and stood in amazement to see the ship, cow and all rise slowly and sail off, disappearing in the north-west.

We went home, but I was so frightened I could not sleep. Rising early Tuesday morning, I mounted my horse and started out, hoping to find some trace of my cow. This I failed to do, but coming back to Leroy in the evening, found that Lank Thomas, who lives in Coffey county, about three or four miles west of Leroy, had found the hide, legs and head in his field that day. He, thinking someone had butchered a stolen beast and thrown the hide away, had brought it to town for identification, but was greatly mystified in not being able to find a track of any kind on soft ground.

After identifying the hide by my brand, I went home, but every time I would drop to sleep I would see the cursed thing, with its big lights and hideous people. I don't know whether they are devils or angels, or what, but we all saw them, and my whole family saw the ship, and I don't want any more to do with them.[15]

The *Advocate* said that Hamilton looked as if he had not recovered from the shock, and everyone who heard him discuss the incident "was convinced that he was sincere in every word he uttered."

Hamilton's story has been quoted in virtually every influential book on the subject of UFOs published during the 1960s and 1970s when the 'airship' wave was rediscovered. The most influential accounts were those which appeared in Frank Edward's *Flying Saucers Serious Business*,[16] and Dr Jacques Vallee's *Anatomy of a Phenomenon*,[17] which reproduced the story in full and reprinted the affidavit signed by ten of Hamilton's friends. Other writers, including John Keel in his classic *Operation Trojan Horse*, used the sighting as historical evidence to link UFO occupants with the mysterious disappearance and mutilation of cattle which were at that time plaguing ranchers in the US Midwest. Keel wrote that the case was significant "not only because of the detailed description of the transparency of the object, but because it was the first in a long line of cattle-rustling reports concerning UFOs."[18]

The airship yarn was deflated in 1977 when UFOlogist Jerome Clark, who had long been fascinated by the story, decided to trace Hamilton's relatives. The farmer died in 1912, but his granddaughter was still living in Yates Center. Others were direct relatives of Hamilton's contemporaries. These included Ben Hudson, the son of the former editor of the local newspaper, Ed Hudson. Hamilton's granddaughter, Elizabeth Linde, then aged 72, said although she liked to believe the story was true, they were aware that the whole episode had been concocted by him and Ed Hudson, then editor of the *Farmer's Advocate*, "following a Saturday afternoon pow-wow." Another old lady, a close friend of Hamilton's daughter Nell, told Clark that a number of local men had formed a club which they called "Ananias" (Liar's Club). She added: "They would get together once in a while to see which one could tell the biggest story they'd concocted since their last meeting. Well, to my knowledge, the club soon broke up after the 'airship and cow' story. I guess that one topped them all."[19]

Indiana-based folklorist Eddie Bullard even found a letter from Hamilton printed in a Missouri newspaper just over two weeks after the 'sighting' where he happily admitted the story was a hoax. Asked about the truthfulness of the story by the editor of the *Atchinson County Mail*, Hamilton wrote: "… I will freely confess to you that I lied about it. I am sure, now, that it was a lightning bug with a cow hair in one foot and an ant crawling up the hair. I must have taken the lightning bug for the airship, the hair for the cable and the ant for the cow. The whole thing must have been an optical illusion."[20]

The tongue-in-cheek amusement which is implicit from Hudson's account demonstrates the lively imagination of the correspondents and staff writers who dreamed up many of the accounts of visitations by the phantom airship in 1897. Despite the public revelation of the hoax, the 'airship and cow' tale continues to appear in popular UFO books to this day, demonstrating both the acceptance of sloppy research in the writings of UFOlogists and the continual reproduction of the story into the mythology of the subject.

For example, as late as 1991 UFO expert John Spencer was still reproducing the yarn in his *UFO Encyclopedia*, more than a decade after the truth about the hoax had been revealed.[21] This burgeoning mythology has led some of the more perceptive writers in this field to realise that 'hoax' cases are as much a part of the phenomenon as 'real' cases. Indeed, the boundary between the two has become increasingly blurred. As Fortean writer Bob Rickard observed, ". . . whether the original incident happened or not, it was published as a true account and possibly thousands of people world-wide have since believed it actually took place. This invests it with a phenomenal reality, and, hoax or not, it conforms to a classical scenario, both ancient and modern, in myth and in fact, for animal kidnaps and mutilations."[22]

UFO historians such as Jerome Clark have argued that the discovery of one clearly identified hoax account should not imply that all the others published in 1896–97 were necessarily bogus. Clark claims a small percentage of the airship stories do suggest that an unexplained UFO-type phenomenon was being described in some cases within the cultural parameters of a time before ordinary Americans were familiar with aerial flying machines. He concludes that "it is difficult to believe that even an amused and hugely tolerant newspaper readership would have put up indefinitely with wholly false reports of a wholly bogus phenomenon."[23] However, the difficulties involved in separating fact and fiction in the published accounts have been outlined in accounts such as the Kansas calfnapping.

The Aurora airship crash

These problems are multiplied in the 1897 saga, where the participants are long dead, and the lack of objective investigations at the time make any conclusive interpretation of sightings impossible. A few key cases, such as the calfnapping hoax, can tell us much about the direction in which truth may ultimately lie, but there is another 'airship' saga which provides an even deeper and more symbolic connection with the modern UFO mythology.

Myth, fantasy and the popular imagination all merge together in the story of the Aurora, Texas, airship crash of 1897, which contains within itself all the elements of a modern fairy tale. Claims about crashed UFOs have become a major industry since the revival of interest in the story that the US Army had captured a 'flying disc' and its alien occupants on a ranch in Roswell, New Mexico, in 1947. These narratives, which were never taken seriously before the 1970s, came half a century after the Aurora crash, which one writer has described as "an old myth that is also new, and in some ways it is Roswell's stillborn twin."[24]

It is worth reproducing the full story as it appeared as a dispatch from a newspaper correspondent in the columns of the *Dallas Morning News* on April 19, 1897:

Aurora, Wise Co., Tex., April 17. – (To the News) – About 6 o'clock this morning, the early risers of Aurora were astonished at the sudden appearance of the airship which has been sailing through the country. It was travelling due north, and much nearer the earth than ever before. Evidently some of the machinery was out of order, for it was making a speed of only ten or twelve miles an hour and gradually settling toward the earth. It sailed directly over the public square, and when it reached the north part of town collided with the tower of Judge Proctor's windmill and went to pieces with a terrific explosion, scattering debris over several acres of ground, wrecking the windmill and water tank, and destroying the judge's flower garden.

The pilot of the ship is supposed to have been the only one on board, and while his remains are badly disfigured, enough of the original has been picked up to show that he was not an inhabitant of this world. Mr T. J. Weems, the United States signals service officer at this place and an authority on astronomy, gives it as his opinion that he was a native of the planet Mars. Papers found on his person – evidently the records of his travels – are written in some unknown hieroglyphics, and cannot be deciphered.

The ship was too badly wrecked to form any conclusions as to its construction or motive power. It was built of an unknown metal, resembling somewhat a mixture of aluminium and silver, and it must have weighed several tons. The town is full of people to-day who are viewing the wreck and gathering pieces of the strange metal from the debris. The pilot's funeral will take place at noon to-morrow. S. E. HAYDON.[25]

The story from Aurora appeared in the *News* not as a world-changing exclusive, but almost at the completion of a long list of dispatches reporting airship 'sightings' from all quarters of the Lone Star State. As Jerome Clark noted, there was absolutely no evidence that anyone at the time took this story seriously. The Dallas paper treated the airship sightings as one huge joke, with editorials humorously proclaiming that the ship came from a secret base at the North Pole. Other dispatches printed days earlier described airships resembling giant Chinese dragons, and another which sailed over Waxahachie seemed to contain "a woman who was running a sewing machine."[26]

If more evidence of the hoax status of the story was necessary, the lack of any contemporary follow-up by Texas journalists in the aftermath of the claimed crash-landing of an alien craft just 45 miles north of the city of Dallas should have provided sufficient proof. But the long forgotten story was revived 70 years later for a modern audience when a copy of the original cutting was sent to a *Dallas Morning News* columnist, Frank Tolbert.

The USA was in the grip of one of the biggest UFO waves of the century in January of that year, and the cutting fuelled Tolbert's interest in the Texas airship wave of the 1890s. His inquiries with retired railroad operators who remembered that era uncovered the sobering fact that the whole scenario had been a well-planned hoax triggered by radio operators and encouraged by a host of jokers across the nation.[27]

Tolbert's article aroused so much interest that Dr J. Allen Hynek, the former scientific consultant to the US Air Force's 20-year-old government-funded UFO Project Blue Book, sent a investigator called William Driskell to Aurora. He discovered that although a Judge J. S. Proctor had lived in the Aurora area, there was never a windmill at his farm. It emerged that the T. J. Weems named in the 1897 account as a military signals officer, did exist but was actually the village blacksmith!

Driskell concluded that the original account was a newspaper hoax, and suggested that the most likely explanation was that Heydon, a local cotton buyer, had invented the tale as a joke. One of the old-timers told him: "... Aurora was a busy little town until the railroads put down their new tracks and neglected to include Aurora in their plans ... Heydon was a writer who lived in Aurora and wanted to do something to help keep people in town and to make it a tourist attraction."[28]

Driskell's conclusions were backed by Wise County historian Etta Pegues, who found that old-timers dismissed the whole story as a hoax. Pegues concluded that if the account had been fact rather than fiction, the pioneer history of the county published in 1907 "would have sold a billion copies."[29] Unfortunately, it did not even mention the 'airship crash'!

Despite these revelations, the story simply refused to go away. In 1973 during another UFO wave, Bill Case, a writer on the staff of the *Dallas Times Herald*, again revived the yarn in a series of articles, and returned to Aurora in a determined effort to find proof of the original story. Case even traced three old-timers, who he said remembered hearing about the airship crash from friends and relatives. One of these was Mary Evans, a 91-year-old woman who was fifteen at the time of the airship crash. She said her mother and father had not allowed her to go to the site, but told her they had seen the pilot's body and it was that of a "small man." Later, she informed other reporters she had been misquoted, and added, "I didn't say it that way."

Another was a G. C. Curley, a "remarkably alert" 98-year-old resident of a nursing home in Lewisville. He said two friends had ridden to Aurora to see the wreck and recalled: "They told me the airship had been seen coming from the direction of Dallas the day before and had been sighted in the area. But no one knew what it was. They said it hit something near Judge Proctor's well. The airship was destroyed and the pilot in it was badly torn up. My friends said there was a big crowd of sightseers who were picking up pieces of the exploded airship. But no one could identify the metal it was made of. We didn't have metal like that in America at that time. And they said it was difficult to describe the pilot. They saw only a torn up body. They didn't say people were frightened by the crash. They just couldn't understand what it was."[30]

Furthermore, Case managed to track down the site of the 'airship crash' on the old Proctor farm, now owned by Brawley Oates. In 1967, Alfred Kraus of West Texas State University searched this site with a metal detector, but found only old stove lids, bridles and 1930s licence plates. These facts were ignored in 1973 when it was claimed that Oates had located quantities of an unusual, silvery metal while cleaning out an old well on his property. He also told a newspaper that he had been receiving phone calls from the Army and the CIA expressing an interest in the 'metal fragments' he found.

These were immediately sent for analysis at three separate laboratories, including the university, an aircraft company and a research institute in Canada. On March 31, 1973, Case was able to claim that the samples had been described as "highly unusual" by two out of three laboratories who were asked to test them. Physicist Dr Tom Gray, of the North Texas State University, who analysed one group of samples, said the pieces contained 75 per cent iron and 25 per cent zinc, along with other trace elements. Dr Gray added that the physics of one piece were puzzling, but overall "they have properties and content common to metals of this area." Although the combination of elements could not have been produced in 1897, it could have been done so at a later date this century.

Dr Gray was initially quoted as saying the physics of one sample "stirs my curiosity as a scientist," but was later forced to emphasise that "he did not mean his comments to indicate the sample was of earthly or extraterrestrial origin."[31] Later analysis found the metal was nothing more unearthly than aluminium alloy used to make pots and pans in the 1920s. The reports of this 'discovery' caused a media sensation and triggered a chain of events which led to a virtual siege of the little town by UFO enthusiasts, reporters and TV camera crews. A *Dallas Times Herald* report of the time read, "Using metal detectors, reporters and UFO investigators have located a remote grave in the country cemetery from which they receive the same decibel readings as they did from the sample of metal Dr Gray, and the aircraft company scientists say they find 'puzzling'."[32]

Caught in the middle of a media scrum, Aurora's few remaining residents were determined to stop outsiders desecrating their family plots and even took to holding nightly vigils to prevent anyone digging up graves in the town cemetery. The main protagonists in the media circus were Heyden Hewes of the Oklahoma-based International UFO Bureau (IUFOB) and Walt Andrus, head of the Mutual UFO Network (MUFON), who fought battles between themselves to reach the grave of the airship pilot first.

Bolstered by Case's articles, Hewes approached the Wise County Court for permission to exhume the body of the pilot from the Aurora cemetery. The cemetery was by this time besieged by representatives of both groups, who proceeded to use metal detectors to sweep the graves for signs of any unusual metals. Soon more than 20 graves in the little cemetery had been disturbed as a result of the publicity. By March that year, the town's Cemetery Association was able to prove a triangular stone which had been identified as the 'pilot's grave' was, in fact, a plot belonging to a local family, and won a legal action to prevent any further desecration. They said they would use "whatever means are available" to prevent any further damage. This was too late to prevent the stone, said to have been inscribed with a crude cigar-shaped design, from being stolen from the supposed gravesite. It has never been returned.[33]

Following the success of the legal injunction, the UFO investigators soon gave up their quest to find the body of the pilot and moved on to fight new battles. Heyden Hewes proclaimed the case a hoax. Bill Case continued his investigation, and a month before his death in 1974, tried to make final arrangements for a subsurface radar scan of what he believed to be the pilot's grave. Since that time, the Aurora case has passed into the UFO folklore, and even the appearance of a low-budget movie of the saga (titled *Aurora Encounter*) in 1985 failed to revive interest in the case. When the film was released, a New York reporter visited Aurora and was told the very mention of the subject sent residents "into profound depression" such was the impact UFO believers had made on their little town.[34] As a result, the lessons it should have taught researchers have been drowned out by the controversy surrounding the Roswell incident and other alleged 'UFO crash retrievals.'

Even today, a small number of UFOlogists continue to believe a question mark still hangs over the reality status of the Aurora airship saga. In 1975 when historian Etta Pegues penned an account of the little town which the railroad forgot, it was painfully obvious the great airship hoax was the only historical byline likely to keep the name of Aurora in the history books. She had to concede that S. E. Haydon deserved due credit for his imaginative piece of writing, which she said was "a beautiful piece of fiction."[35]

A British UFO researcher wrote to Aurora Town Council in the 1970s asking if there was any truth to the story and, if so, whether the location of the pilot's burial had been preserved to this day. In a straightforward reply, the town clerk responded: "I regret to say that the entire event was a hoax. This statement I base on the lack of evidence to support such a happening. I have read much about the event and most written accounts of what happened contain very little evidence to support the claims that an airborne craft crashed or even appeared anywhere near Aurora. More important than this, however, in talking with citizens of the community who were alive then, I find that all of them report the entire thing as a hoax … No record exists to support the claims that the pilot was buried in the cemetery here, and it only seems logical that if such an event did take place, it would have received extra attention in the burial plot records. Also, no gravestone exists that would indicate that there was anyone unusual buried there."[36]

Like other folktales, the Aurora crash contains many of the key elements of myth found in the UFO folklore. The symbols which occur in the story were based upon accounts the author had read in the Texas papers during the airship wave. Elsewhere during the wave there were other accounts describing crashed airships which sound very familiar today. In Iowa, a large search was organised after the airship was seen plunging into a reservoir, while a wrecked craft which

THE ROSWELL CRASH

A terrible storm struck desert scrub between Roswell and Corona, New Mexico, on July 4, 1947. Rancher William Brazel heard an explosion and decided to check his land. He was amazed to find debris – tough but flexible metal foil and pieces of what resembled balsa wood, stained with purple/pink ideograms – scattered over a wide area.

Bemused, he collected some wreckage and considered options. One neighbour suggested a "balloon bomb," Japanese balloons strung with high explosive that crossed the Pacific in their thousands during the Second World War.

Flying saucers had just made headlines, so Brazel took debris to Roswell Air Force Base on his next trip to town. Major Jesse Marcel, an intelligence officer, collected all the wreckage in a pick-up. A Press release from the base swiftly announced it was a crashed flying disc. This brought massive media interest, until the Pentagon ordered Roswell into silence.

Marcel flew with the debris to Carswell Air Base in Texas en route to Wright Field in Dayton, Ohio, where "foreign technology" was taken to be studied. But at Carswell, General Roger Ramey advised that the debris was just a weather balloon, the Press were called and an embarrassed Marcel was pictured holding balloon remains to kill the story.

Thirty years later, on retirement Marcel spoke to UFOlogists. He claimed the balloon story was a ruse to silence the Press. The wreckage was flown on to Dayton and thought to be an alien craft. As the Roswell bandwagon has grown, many people have come forward to say they saw debris or a crashed disc elsewhere in the desert. Others claim small alien bodies were recovered and taken to the base.

After many movies, books and TV dramas such as *Dark Skies* and *The X Files*, all post-1977 before which the case was unknown, Roswell is no longer just the world's best-promoted UFO case, but a legend in its own lifetime.

Is it real?
● Marcel appeared to be a sincere witness with no axe to grind, reporting what he believed. He stuck by his story until his death, insisting this case proved to the USAF that aliens were here.
● The US government agrees that the incident did occur and was first believed to be a UFO by personnel at Roswell. Records – such as internal memos and FBI data – exist discussing it.
● Several witnesses who saw the wreckage describe its unusual "unbendable" properties.

turned up near San Francisco turned out to have been pulled there on the back of a wagon. There were even hints of sinister Government cover-ups, with claims that airships were being constructed at US military installations, such as Fort Sheridan near Chicago and Fort Logan in Colorado. One report even claimed, "A profound secrecy has been maintained as to what has been accomplished, even army officers themselves only getting vague inklings of what is going on."[37]

The craze for seeing airships in the US during the 1890s was the product of a culture obsessed with science and flying machines. Today in an age preoccupied with space travel we see flying saucers and alien spaceships. But the same mythological elements present in the airship sightings also turn up in the later stories surrounding crashed saucers and pickled aliens hidden inside Air

Is it solved?
- The debris was found close to the location of much secret research into nuclear weapons and space flight technology. Roswell base was, in fact, the only one then equipped with a nuclear arsenal.
- A highly secret project, code name Mogul, operated locally. This involved specially adapted balloons sent high into the atmosphere to search for evidence that the USSR had exploded an atom bomb. The project was strictly 'need to know.' Most people at Roswell had no need to know.
- The scientist who masterminded Mogul talked openly only in 1995. He has studied the Roswell case against his balloon launches and claims they lost a test device in this area around the right time. The balloons did use very similar equipment to that found at Roswell, including tape with the same colours and flower image patterns as reported by Brazel.
- Major Marcel's son, now a doctor, saw the wreckage and was recently regressed back to that day by a psychiatrist. His account offers a description of what sounds like wood, metal and Bakelite, materials used by Mogul, but unlikely candidates for the construction of an alien spacecraft.
- An investigation by the GAO (General Accounting Office) after senatorial pressure in 1995 was given free access to US government archives seeking data. It found only clues suggestive that Mogul was the culprit and that the US government invented the "weather balloon" story to prevent the truth from emerging, probably without even telling Major Marcel he was being used.

Conclusion
This is one of the world's most highly touted UFO cases, but events since 1995 have seen it disintegrate. On balance, there is now every reason to accept that the wreckage was indeed from a balloon, but not an ordinary weather balloon as the US government claimed at the time.

A covert experiment was hidden and then lost amidst Cold War bureaucracy allowing the paranoia about UFOs to take over 30 years later. Many UFOlogists still believe Roswell is a genuine alien UFO crash. Whilst a few unanswered questions remain, several of the witnesses making the more extreme claims (such as seeing alien bodies) have provided testimony that has not stood up to scrutiny. It is hard to study this case objectively and regard it unsolved.

Further reading.
The Roswell File. Shawcross, T. Bloomsbury, London, 1998.

Force bases which are a persistent feature of UFO folklore at the end of the 20th century.

The crash of a craft in a remote desert location, the small stature of the alien pilot, the discovery of strange metal at the crash site, the presence of 'hiero-glyphs' on the wreckage, and the willingness of 'eye-witnesses' to claim involvement in events which never occurred are all present at Aurora. The tale was merely the forerunner of that great-granddaddy of all flying saucer yarns, the Roswell incident of 1947. Imagination, fantasy and wish-fulfilment created the Roswell story, but these all had a long and distinguished pedigree in the fairytale which was told by a cotton buyer once upon a time in the west.

2

THE HOWDEN MOOR INCIDENT

David Clarke

"During a series of spectacular UFO encounters over central England ... six Tornado, RAF jets were in pursuit of a 300-foot Flying Triangle UFO, which hovered at times within feet of houses in Dronfield, Sheffield. The British Geological Survey instrumentation at three sites on March 24, 1997, recorded two air burst explosions. It is believed a UFO conducted a hostile attack on aircraft that were in pursuit. Apparently, one RAF jet was lost and two bodies were recovered from a mountain reservoir the next morning. All air traffic was excluded from the area." – Internet posting, *UFO Updates*, March 4, 1998.

One of the first lessons to be learned by those who venture into the weird world of UFOlogy is that the most mundane of events can be misconstrued and misinterpreted by those who are actively engaged in the creation of mythology. For believers in crashed UFOs, government cover-ups and alien abductions, everyday events take on an entirely new meaning and significance, and mysteries can only be explained by reference to the most fantastic and paranoid of theories. All these influences can be seen at work in the Howden Moor incident, which at the time of writing is two years old and yet has already been proclaimed by some as Britain's answer to the Roswell flying saucer crash. In addition, its promoters have produced their own self-perpetuating literature whose narrative follows the form and substance of other classic 'crashed saucer' folklore.

Britain has produced few, if any, cases which could be claimed as classic 'UFO crash retrievals.' For some years there has been a desperate hunt by believers to locate a story which could be trumpeted as an equivalent to the legendary Roswell incident. UFO researcher Nick Redfern has claimed there is evidence of a British equivalent of the fabled MJ-12 group of high level military intelligence officials who are actively engaged in the cover-up of frequent ET crashes on earth.[1]

Claims that UFOs crashed on British soil as early as the 1940s and have been spirited away by the military are common in UFOlogy, but the proof is always based upon selective use of 'evidence,' misinterpretation, wishful thinking and the unsubstantiated claims of 'anonymous' informants. Sociologists have observed that crashed UFO narratives are created as a result of a desire to satiate deep psychological needs among those who believe them. On another level, they are the raw material of many lucrative books and TV programmes for those who cater for such needs by promoting these stories.[2]

To create a crashed UFO narrative, a vital ingredient is a charismatic and proactive individual who can promote the story and convince other like-minded people that a UFO did crash, and that a gigantic conspiracy is underway to conceal the facts of the case. Once this level of unquestioning belief is established, 'facts' soon become distorted by the will to believe. Any suggestion of a rational explanation will be seen as a product of the inevitable cover-up which this mindset expects to find.

The Howden UFO conspiracy could have been lifted straight out of a plot from *The X Files* TV series: an unidentified flying object hovering in the clear night sky; callers jamming police switchboards to report an object on a collision course with the Peak District hills; RAF jets screaming through the sky as if in pursuit of something, and finally a deafening explosion which sent gamekeepers rushing from their isolated cottage on a remote moor, triggering a massive search operation by the authorities. That is a summary of the events of Monday, March 24, 1997, over the Howden Moors, a wild and hilly region on the border between the Yorkshire city of Sheffield and the Peak District National Park.

This kind of location is ideal territory for stories of a UFO crash to germinate. It is one of the most remote regions of central England, consisting of high moorlands dotted with huge man-made dams used as a training ground by the famous 617 Dambusters Squadron in the 1940s. It is also a region with an extensive history of aircrash tragedies, with over 50 planes lost on the Dark Peak moors since the Second World War. These events have helped to create an extensive folklore which has grown to produce a local legend that on certain moonlit nights a 'ghost plane' can be seen skimming the still waters of the Derwent Dams.[3]

It is this background against which the sightings of that evening have to be set. The sequence of events which transpired that night have led some UFOlogists to conclude that a massive cover-up is underway to hide the truth from the public concerning a potentially earth-shattering event, a hostile encounter between extraterrestrials and NATO air defence forces. Other more rational observers have concluded that this 'incident' was simply the product of a series of unconnected events and coincidences arising from a routine military exercise, coupled with the misperception of natural and man-made aerial objects. The case may well have had a mundane origin, but the mystery led to questions in the British Houses of Parliament and the creation of one of the biggest UFO controversies ever seen in Britain, set to be perpetuated for years to come in books, magazines and on the Internet.

Whatever the source of the phenomena, it is clear that the authorities took the initial reports seriously. For within minutes of the first calls that evening, the emergency services mounted one of the region's largest ever air and ground rescue operations in response to a suspected air disaster, an operation which involved almost 200 personnel and cost in excess of £50,000. For 15 hours, emergency services from four different counties, co-ordinated by South Yorkshire Police, were involved in searching up to 50 square miles of the most barren Pennine moorland for wreckage from a suspected aircrash, a crash which we now know never happened. This account aims to clear away the aura of mystery which has continued to surround these events, cutting through wild speculation to reveal the truth. For while the Howden Moor incident was not initially a UFO case, it contains all the hallmarks of one and has been promoted as such more recently by those who today have become the self-styled proponents of modern belief in alien visitations.

The aircrash which never was

It is a story which begins on a clear, cold spring night in the Peak District National Park, a night when many necks were turned skywards in search of the spectacular Hale-Bopp comet which at that time was prominently visible. The tranquillity of that evening was shattered shortly after 10 pm when the operations room at Ecclesfield Police Station in Sheffield began to receive emergency calls from people living in the area of Bolsterstone, an isolated village high on the moorland border between the city and the Peak National Park.[4]

The first call came at 10.15 pm from farmers near Bolsterstone, who asked the control room staff if they had received any reports of aircraft coming down over the moors. They said they saw a plane flying low and disappearing over the highest point on the western horizon, formed by the moors known as Featherbed Moss, followed by a "flash," plumes of smoke and an eerie red glow.

Shortly afterwards, further calls were received both by South Yorkshire and the Derbyshire forces, reporting an "aircraft in distress" and "a plane having gone down west of the Midhope Moors area." These circumstantial reports were given additional substance from gamekeepers at Strines Forest, who told of hearing an enormous explosion and seeing a "large orange glow" visible on the horizon.

At 10.30 pm, with a number of consistent and reliable reports at hand, an alarmed police controller called out 40 police officers and placed the county's Fire and Rescue Service and Ambulance Service on full alert in anticipation of a possible disaster involving a light aircraft. By 11 pm, Chief Inspector Christine Burbeary had taken control of the incident and asked for neighbouring West Yorkshire's police helicopter to be scrambled to help the search for the aircraft. From the beginning, it became clear this was no ordinary sighting of a plane in distress. Contact with Manchester Airport ascertained that no emergency calls had been received or aircraft reported missing. Furthermore, nothing was registered on the airport's radar, which covers a large segment of the northern Peak District. Staff in both Sheffield and Derbyshire were now placing urgent calls to civilian and military airports who might have had traffic flying above the Peak District. However, the message that came back from the RAF was loud and clear – "It is not one of ours."

By 11 pm, West Yorkshire Police's helicopter had reached the moors near Bolsterstone and was beginning a large-scale search of the area using its searchlight and thermal imaging equipment to detect signs of a fire or wreckage. It was joined at midnight by a Sea King helicopter from RAF Leconfield on the East Coast, which constituted the single military input to the search operation. Use of the Sea King had been authorised by a Flight Lieutenant at RAF Kinloss in Scotland, a base which co-ordinates air–sea rescue operations around the coastline of north Britain.

A check by Kinloss on radar information for the South Yorkshire area and surroundings found nothing significant at the time of the incident, according to South Yorkshire Police's log of the incident.[5] However, when Kinloss ran a check with the British Geological Survey in Edinburgh they found a sonic boom had been recorded in the Sheffield area which coincided with the reports of an explosion reported by witnesses. However, the signal recorded was not consistent with an object crashing into the ground, but rather of a mid-air explosion which sent out pressure waves, such as would be expected in the wake of a supersonic aircraft.

On the ground, fire and rescue tenders from stations in South Yorkshire were racing towards the moors whilst staff at Sheffield's Royal Hallamshire Hospital were reportedly warned to stand-by to receive possible survivors or casualties from the 'aircrash.' The fire crews rendezvoused at the Strines Inn, a public house which stands in an isolated spot high on the moors. However, with no clear information concerning where the 'crash' was located and the possibility that it could have occurred anywhere in a wild and inhospitable zone, police and fire crews had no option but to call upon the mountain rescue service for assistance. Excitement was caused at one point when the crew of a mobile police patrol spotted what they thought was "a plume of smoke" rising into the air from one part of the suspected crash zone. The police helicopter crew were asked to investigate and shine their spotlight on the source, but it was later concluded that the "plume of smoke" came from a cement factory some distance away on the horizon![6]

By midnight, dozens of volunteers from the seven mountain rescue teams in the Peak District were being contacted by phone, some roused from beds, others asked to leave their places of work and join the search operation. The Peak District's Mountain Rescue Service (PDMRS) established their headquarters at the service's Hepshaw Farm base on Langsett Moor. The farm was later used as a rendezvous base for the Sea King, which landed there on several occasions during the course of the following 12 hours, picking up mountain rescue staff and equipment to help the search.

By the early hours of March 25, a total of 141 volunteers from all seven Peak District teams were out on foot searching the difficult and often treacherous terrain stretching from Broomhead Moor, west of Bolsterstone, out towards the vast and uninhabited tracts of peat bog north of the Howden Reservoir complex. Joining them were police units and dog teams from the Search and Rescue Dog Association. The PDMRS commanders split this large group of personnel into groups, each assigned sectors of the moors to search on foot with help from dog teams. This was a long and painstaking process, but resulted in a thorough search which was able to rule out any chance of overlooking a crash site. Believers in a UFO-related crash were later to claim that the highly experienced teams had been directed to check the wrong area of moor while the 'real' search operation was underway by the military elsewhere. However, the promoter of this theory was forced to admit he had never spoken to any of the mountain rescue crews to ascertain the facts![7]

Before the full ground search got underway, the West Yorkshire helicopter found no evidence to indicate an aircrash of any kind had actually taken place. But calls were continuing to police stations both in South Yorkshire and Derbyshire reporting a low-flying aircraft, an explosion and a flash over the moors. One of these corroborating reports came in shortly after 1 am from a police Special Constable, who saw what she thought was a light aircraft flying very low and apparently on a collision course with the moors while she was driving near Bolsterstone around 10 pm.

Marie-France Tattersfield had, like many others, gone out to take pictures of the spectacular Hale-Bopp comet which was clearly visible in the evening sky. She was driving near the village of Bolsterstone with her pilot husband Steve when they were suddenly shocked to see what looked like a large four-seater aircraft flying directly across their field of vision at an alarmingly low altitude.

"It was the weirdest thing I've ever seen," she said. "... It was a big aeroplane and was well below the legal altitude for safe flying. All its windows were lit up, which made it look even more odd as no light aircraft pilot would fly blind at that time of night over these hills."[8]

The couple held their breath as the aircraft disappeared in the west towards Broomhead Moor, but it was the call from gamekeeper Mike Ellison and his wife Barbara which left the night inspector in no doubt an aircrash of some kind had occurred. The couple live in an isolated cottage on the moors at Strines, on the border between South Yorkshire and Derbyshire. Shortly after 10 pm that night, they were watching TV when they were shocked to hear a massive explosion – "A boom and bang in the sky outside."[9] The pair both rushed outside, expecting to find an aircraft had crashed onto the moors or into the nearby Strines reservoir, but saw nothing except an eerie glow over the distant moors. Early the following morning while the search of the moors was still underway, South Yorkshire police set up a special phone line for the public to report sightings of the mystery aircraft. They were immediately inundated with information from people who had seen low-flying aircraft, military jets and UFOs across a wide area, stretching from Chesterfield in the south to Thurgoland, thirty miles away on the border between South and West Yorkshire.[10]

One of these additional witnesses was an elderly lady in Stannington, several miles to the south of the 'crash zone,' who saw a strange object from her bedroom window just seconds before Marie-France Tattersfield was buzzed by the strange aircraft. This 81-year-old pensioner was watching the comet from her bedroom window facing the moors when "I saw what I first thought was a plane come over the top of the hills beyond High Bradfield. It went towards Strines in the west and was shaped liked a long cigar, which looked as if it was on fire because it glowed. I couldn't make out any wings and it made no noise at all. The light just glowed, it didn't flash, and it was very queer looking."[11]

This UFO was flying at rooftop level and disappeared quickly towards the hills beyond Bolsterstone. But the most important sighting reported that night came from gamekeeper John Littlewood, who saw an object pass over his head whilst working on Midhope Moor, near Bolsterstone, just after 10 pm. Mr Littlewood was in his four-wheel drive vehicle when he saw two red lights in the sky arriving from the direction of Stocksbridge. As he watched, the lights approached, and it became obvious they were attached to the wings of an aircraft.

"It was definitely a plane, and it was a big one," said Mr Littlewood. "It was like an old-time plane, but different to a Lancaster ... it came right over the top of me. I could see there were no lights on it like you usually see on aircraft, just two red ones. It was making a loud humming noise and disappeared towards Woodhead. It was very long, slow and low, probably about 500 feet in altitude. It was a clear night and I could clearly see the outline, and it was making quite a noise. I just thought 'What the hell is that?' and took it to be a military plane, but I could not understand why it had got just the two lights on its wings. The planes I regularly see have the usual flashing strobes, but this one was different. Later, when I got home, I found my neighbour's little girl saw the lights through the bay window of their house and had run out to tell her dad she thought she had seen a flying saucer."[12]

Chief Inspector Burbeary said these later reports simply served to confirm the earlier information that a plane had actually gone down on the moors. Despite scepticism from her opposite number in the Derbyshire police, who refused to order his officers to join a similar search operation, she decided to scale up the investigation in the early hours and called in additional mountain rescue volunteers. She said: "My concern was that we could have people from a crashed aircraft lying on the moor seriously injured. It was an exceedingly cold night and we had to find them straightaway."[13]

By 7 am the following morning, the RAF, after consulting with the Civil Aviation Authority, authorised the setting-up of what it called a "Dangerous Flying Zone" with a ten-mile radius, centred upon the Howden Reservoir from where the ground search was directed. As a result, Air Traffic Control staff at Manchester were notified and airliners stacking up at high altitude warned of the flying restrictions below their flight corridors. The Danger Zone was established, as later admitted in Parliament, as a routine measure to allow the two helicopters to complete their search unhindered by both military and civilian aircraft, particularly TV camera crews.

In the event, as dawn broke the two helicopters made further thorough checks of the moors, maintaining radio contact throughout with volunteers searching on the ground, without finding anything. The Sea King returned to its base at Leconfield at 2 pm, as Chief Inspector Burbeary gave the order to scale down the operation. She admitted: "We got nothing back from air traffic control, no reports of aircraft failing to return. Eventually, having looked at all the circumstances, the decision had to be made to call the search off. The conclusion at the end of the search had to be that no aircraft crashed on the moor."

However, the experienced senior officer was forced to admit that there *had* been a number of genuine reports of 'phenomena,' including a low-flying aircraft, a huge explosion in the sky, and smoke. Mike France, the mountain rescue co-ordinator, remained as baffled as the police. He said his teams, who knew the area intimately, had thoroughly searched up to 50 square miles of the moorland with help from the helicopters. "There was no scouring to the moor," he concluded. "There were no bits of wreckage. There were no oil traces on the reservoirs."

Ghostplane or UFO crash?

Officially, South Yorkshire Police classified the incident as "unexplained," but senior officers remain convinced that the truth about what really happened has still to be revealed. Several have admitted they suspect the incident was at least in part triggered by a covert operation involving a military aircraft or a remote-controlled aircraft.[14] At the time, the closing entry in the Major Incident Log opened by the police suggested it had been caused by a series of unconnected events and concluded, "Enquiries reveal a combination of circumstances that would lead people to believe a plane might have crashed."[15] Today, the force continues to remain open-minded about the source of the events and have even considered the possibility that the search was triggered by a plane involved in an illegal drugs drop or the appearance of the 'ghost flier' which, it is claimed, haunts the moors surrounding the Ladybower and Derwent reservoirs.

Since that time, two TV documentaries have discussed the case and there has been a flurry of speculation linking it with UFOs and sinister government cover-ups. To the police, emergency services and local people who took part in the events, the Howden Moor incident remains a baffling mystery, but one of mundane proportions, and none of them has ever seriously considered the involvement of alien visitors in the search for the crashed plane. However, the weeks which followed the incident saw an influx of alien hunters and UFO enthusiasts who visited the search zone looking for evidence of a crashed saucer, quizzing bemused residents and farmers for 'evidence' which might confirm their beliefs. It was not long before rumours were circulating concerning mysterious burnt patches in fields, 'Men in Black,' flying triangles, objects removed from the moors on low-loaders and unmarked helicopters.[16] The case may never have been transformed from a mundane hunt for a 'crashed aircraft' to a UFO cover-up were it not for the influence of one person who is entirely responsible for the promotion of the myth.

UFO investigator Max Burns makes no secret of his belief in aliens and a huge cover-up by the authorities. In a posting on the Internet one year afterwards, he proclaimed the case as "One of the biggest UFO incidents in recent years involving a huge flying triangle, military jets, sonic booms, a bolide meteor, unmarked helicopters, glowing orange objects and what I hope, when you have studied the evidence, you will agree is a conspiracy on behalf of the civilian and military authorities to hide the facts from the public."[17]

Burns contends that the case arose out of an incident whereby RAF jet interceptors were scrambled to pursue a slow-moving triangular UFO of extraterrestrial origin over Sheffield and the Peak District. His evidence included the statement of an "anonymous" Royal Signals radar operator who, through a third party, claimed to have detected a UFO on a screen at RAF Linton-upon-Ouse at 9.55 that night. He was later to claim he was warned by his superiors not to discuss the case.[18] As a result of this imagined chase, one of the pursuing jets was attacked and destroyed by the UFO in Burns' scenario, subsequently crashing to earth onto the moors, or into one of the reservoirs, with the loss of at least one of its crew members.

The sonic booms which were recorded by the British Geological Survey that night were to Burns proof of an 'air burst' caused by a jet exploding or an ET weapon firing upon pursuing jets. Furthermore, a report by a passenger in a minibus was made to the police describing a dark-skinned man wandering on the A57 road at the Ladybower Viaduct apparently covered in fuel. This was one hour after the reports of the aircrash and has been interpreted by Burns as a 'sighting' of the co-pilot of the downed Tornado, who was desperately trying to hitch a ride back to the nearest city to alert the military! As a result, Burns felt he was able to conclude that the man spotted that night was the pilot or co-pilot of the Tornado jet he believes was lost above the Peak District, an incident which he alleges is the subject of a massive and high-level cover-up and 'dirty tricks campaign.' To Burns, rational explanations put forward later were 'cover stories' deliberately invented to conceal the tracks of the military retrieval squads. He wrote:

> It is without doubt that the military are involved in a large cover-up regarding the attempted interception of the triangle [UFO], including conspiracy, the placement of cover-stories [and the] debunking of witnesses ... However, [there is] damning evidence from a [witness] who encountered the pilot or the co-pilot on the Snake Pass about an hour after the explosions occurred, stinking of aviation fuel within three miles of Howden Moor ... The distance of three miles from the 'crash' site had been covered in one hour. Allowing for the time necessary to unhitch his parachute, then walking at 4 miles an hour would place him exactly where he was encountered by the mini-bus. I feel that this is without doubt the co-pilot of the Tornado jet, who was soaked in aviation fuel and was making his way to the nearest metropolis to alert the military.[19]

The theory that the mystery man soaked in fuel was actually an RAF or NATO pilot clearly stretches credibility to its limits. Nick Pope, the former head of the MoD department at Whitehall which deals with UFOs – he is now a successful author of popular books on the subject – found this claim particularly hard to swallow. He said: "I think it's ridiculous to suggest that this has anything to do with the RAF, on the basis that a pilot from a downed jet would always stay at the crash site, waiting for the inevitable military search and rescue operation. He'd be wearing a distinctive dark green flying suit that I think even a layman would realise was military-issue."[20]

The lack of one single eye-witness reporting such a mid-air encounter between a jet and a UFO, and the improbability that the wreckage left in the wake of a crash would not have been immediately spotted by hundreds of volunteers searching the moors within hours of the incident, did not faze Burns. He accounted for the lack of wreckage by claiming mountain rescue teams had been deliberately misdirected to search the wrong area of moor. When that claim became untenable, he raised the possibility that the downed Tornado dived beneath the waters of one of the 13 reservoirs which litter the high moorland region between Sheffield and Manchester. This was speculation which had been mooted on the night of the incident by fire crews and others, but was quickly ruled out by Yorkshire Water workers, who checked a number of the reservoirs and found no evidence of oil slicks or wreckage which would have been evident had a crash of a 20-ton warplane really occurred.[21]

These conclusions were echoed by Nick Pope, who commented on the "sheer implausibility" of the claims made by those who believed a jet had crashed. He said: "If an RAF aircraft had really crashed (whatever the circumstances), it would have been virtually impossible to implement a successful cover-up. There would almost inevitably have been a fire, and in those circumstances the emergency services and members of the public would have located

THOMAS MANTELL – JET PILOT KILLED BY UFO?

At 1.20 pm on January 7, 1948, Godman Field Airbase began taking calls from state police about an unidentified aerial object moving in from the east. By 1.45 pm, control tower personnel at Godman could also see the object. It was generally agreed to be small and white with a shape like an "ice-cream cone" or an "umbrella."

A flight of four F-51 planes were in the area. Godman control tower contacted their leader, Captain Thomas Mantell, requesting they investigate. As Mantell closed on the object, he was still unable to make a positive identification. His radio communications suggested the object was "metallic" and of a "tremendous size."

Mantell's wingmen dropped out of the chase at 22,500 feet due to lack of oxygen, leaving him

alone "climbing into the sun." A few moments later, his plane plummeted to earth – and Mantell was dead. The headline in that evening's *Louisville Courier* claimed "F-51 and Capt. Mantell Destroyed Chasing Flying Saucer."

Is it real?
● The Air Force never denied the crash took place.
● The 'UFO' was seen by numerous witnesses throughout the day.
● Air Force investigators initially claimed Mantell had mistakenly chased Venus. When this was proved impossible in daylight, flying saucer buffs took it as evidence of a UFO cover-up.

Is it solved?
● Mantell's wingmen saw nothing resembling a flying saucer.
● Later investigation revealed that a classified Navy 'Skyhook' balloon

had been launched a day earlier and crossed the area.
● Several descriptions of the object accurately described a Skyhook balloon.

Conclusion
The Mantell case represents one of UFOlogy's enduring myths. The notion that 'one of ours' was destroyed by, or whilst chasing, 'one of theirs' occurs again and again, most recently in the Howden Moor incident. The facts available suggest it was nothing more than a tragic accident caused by Mantell vainly pursuing a then top-secret balloon.

Further Reading
"Mantell Incident." pp. 351–355. Clark, J. *The UFO Book*. Visible Ink, 1998. *UFO: The Government Files*. pp. 24–26. Brookesmith, P. Blandford, 1996. *Something in the Air*, pp.24–31. Randles, J. Hale, 1998.

the crash site very quickly. The military search and rescue operation would also have been readily apparent, as would the subsequent removal of wreckage ... These are not things that would go unnoticed, especially in a situation where the emergency services were already conducting a search.

"Indeed, the crash of an aircraft into a reservoir would probably make a cover-up more difficult. Not only would oil slicks and surface debris have been visible, but recovery of the wreckage would have taken longer, and couldn't have taken place without the knowledge of personnel from Yorkshire Water. Whichever way you look at this, a considerable number of people from various different agencies would have been involved, or would have heard about or seen some activity. Clearly this is not a situation in which a cover-up could be successfully executed."[22]

In addition, the half-hearted response of the military to the report of the 'aircrash' hardly fits the scenario envisaged by Burns. It is clear from the testimony of local people, the fire service and the civilian mountain rescue staff that other than the Sea King crew no other military personnel were involved in the ground search, as would have been the case had a warplane really been reported missing. One may be tempted to laugh at such a bizarre and extreme claims, but Burns' 'evidence' has been taken seriously by many 'UFOlogists' in the USA and elsewhere who were not aware of the correct sequence of events of that night.

So what did happen? A number of theories were aired in the local and national Press in the days following the crash. Most prominent of these concerned the stories of ghostly aircraft which had become part of the Peakland folklore since the events of the Second World War. Reporters from national newspapers connected these stories with the search and soon transformed the events into the latest sighting of the fabled 'ghostplane' in what a *Daily Express* sub-editor dubbed "the Howden Triangle."[23] There were other theories, too. One of these was that the incident had been caused as a result of a covert drugs drop on the moors, involving a light aircraft whose pilot had been well paid by drugs barons in Manchester or on the Continent to take a dare-devil risk by flying low and undetected by radar. This was the explanation favoured by Mike France and his colleagues in the mountain rescue service, but the police said there was no evidence to support the claim.

Sonic events and military denials

One of the final entries on the police log of the Howden incident concerns the report from the British Geological Survey about the sonic boom reported to

RAF Kinloss. The BGS operate sensitive recording equipment across the British Isles listening out for earth tremors and 'quakes which occasionally pick up other sonic anomalies, including air blasts caused by objects exceeding the speed of sound. When the BGS were contacted later, it was found that their equipment at three separate stations on the Pennines had independently recorded a sonic boom at 10.06 pm, corresponding in time with the reports of the crash and explosion heard near Sheffield. Fourteen minutes earlier, another sonic event had been detected, also from the Sheffield region.[24] These were two of just eight similar sonic events recorded across the whole country that year. Experts at Edinburgh confirmed the culprits were nearly always the same – military aircraft. The only alternative sources of sonic events were Concorde, and on rare occasions space debris such as bolide meteors which burn up in the atmosphere, producing lights in the sky and mini-explosions.[25] There was no evidence in this case to suggest that space debris had been responsible.

In the Howden case, seismologists at the Edinburgh unit were even more specific. They said the two sonic booms could only have been caused by a military aircraft reaching supersonic speed, possibly while performing a mid-air turn during a low-flying exercise. BGS records showed how they had contacted the RAF low-flying complaints department shortly afterwards, and were told the military "could not confirm" one of their jets was responsible for the two breaches of flying law. As the MoD admitted later, the Military Flying Regulations forbid pilots to accelerate through the sound barrier above land, and any breach of the regulations must be reported by pilots to the authorities within minutes of landing. Failure to do so could lead to a police investigation and the possibility of a court martial.

In this case, the RAF had quite clearly stated to both the police and the Press that there was no exercise underway that night, and even went so far as to send a helicopter to help the search effort. This posed a question: would the military have gone so far as to lie just to provide an elaborate cover for a minor, albeit embarrassing, blunder like an aircraft breaking the sound barrier? The apparent improbability of this scenario only served to fuel the suspicions of UFO buffs that there was more to this case than met the eye, and added weight to the mushrooming claims that something really had crashed on those moors.

Suspicions were further aroused by the number of individuals who reported seeing low-flying military jets apparently taking part in an exercise of some kind earlier that night. Between 9.45 and 10 pm, people in Dronfield, a small town in southern Derbyshire thirty miles away from the 'aircrash' zone, reported seeing a huge triangular-shaped UFO. One woman claimed this brightly-lit craft made a humming noise like an electricity sub-station and flew over her home at an altitude of just 300 feet, followed minutes later by pairs of low-flying RAF jets all heading north towards what would later become the search zone. "It was like an airshow," she claimed. ". . . I have never seen as much air traffic in the sky at the same time."[26]

It was this particular sighting which excited the UFO believers the most as it appeared to describe a triangular-shaped UFO being pursued by military aircraft, a scenario which, added to the sonic events recorded minutes later, appeared to support the claim that these were evidence of aircraft exploding or the UFO crashing. On the other hand, different witnesses did see aircraft that same evening, but did not report a huge triangular UFO, which, if the female eyewitness was to be believed, should have been plainly visible to hundreds of people. One of these was an ex-airman, John Brassington, who described how he distinctly heard both a single-engined plane and minutes later two jets – possibly Tornadoes – so low that they shook the foundations of his house. "I can assure you that if the RAF say nothing was going on that night, they are being economical with the truth," he added.[27]

The most convincing evidence of all came from an off-duty police sergeant who saw a formation of low-flying fighter jets pass overhead whilst driving near a village in the northern Peaks. He said it appeared one low-level jet was being pursued by a formation of others as part of the exercise, the whole formation appearing to the eye as a gigantic 'V' or triangle in the sky.[28] A further clue came from a call to the police from a man who noticed a triangular-shaped object hovering in the sky above the Barnsley area whilst on a train at 7.40 that night, more than two hours before the drama began. He said it was covered in lights and was not like any aircraft he had seen before. He told the police control room: "I don't think it's a plane. I'd advise you to call off the search as you're wasting your time."[29]

These reports pose an obvious question. Was a triangular UFO being chased by RAF jets or was it actually part and parcel of the exercise which was patently going on that night? Just one month before these sightings, a major military exercise code-named "Northern Adventure" had taken place over the North Sea. This was an exercise which coincided with a sighting by an aircraft enthusiast of a triangular-shaped object under escort from RAF Tornado jets, flying off the coast at Mablethorpe in Lincolnshire. This witness described the triangle as being around two-thirds the size of the two escort jets, with the

formation heading out to sea at low altitude and at subsonic speeds.[30] The proximity in time between this observation, the North Sea exercise and the sightings
of March 24, 1997, are too close to be unconnected. They suggest the 'triangular object' escorted by jets could have been either a prototype Stealth aircraft
similar to those alleged to be under development by British Aerospace – and
widely believed to be operating from their plant at Warton in Lancashire – or
an Unmanned Aerial Vehicle (UAV) piloted by remote control, which was
taking part in a covert training operation. Mablethorpe is just south of RAF
Donna Nook where target drones used in coastal exercises at this firing range
have been reported as UFOs before.

Having collected all the available evidence surrounding the Howden Moor
incident, I arranged a meeting with Labour MP Helen Jackson, whose Hillsborough constituency includes the aircrash zone. She agreed there was a clear
case for probing questions to be put to both the MoD and the Home Office
concerning the source of the sonic booms and the cause of the sightings which
sparked the futile – and costly – search and rescue operation. Quite apart from
anything else, if military aircraft caused the booms and the search, flying laws
had been broken and public money wasted.

Questions in Parliament

Precisely one year after the Howden Moor incident, Mrs Jackson tabled seven
written questions to the then Defence Minister George Robertson. These
included whether RAF or NATO aircraft were engaged on an exercise over
northern England and what complaints had been received concerning low-
flying aircraft that night. She also asked if the sonic booms detected by the BGS
were caused by aircraft, and if any reported sightings of UFOs had been
received from the public or police.[31]

On March 30, 1998, John Spellar, the Under-Secretary of State for Defence,
admitted that a number of military aircraft were booked to carry out low-flying
training over northern England on the evening of March 24, 1997. This was a
blatant contradiction of statements the MoD made at the time of the search and
rescue operation, but was conveniently accounted for by the claim that the exercise began at 8.30 and was over by 9.30, a full thirty minutes before the
'aircrash' reports. Mr Spellar also confirmed that MoD received 13 complaints
about aircraft activity from a number of locations across the UK, but none of
UFOs! On the subject of the sonic booms, he said there was "no record of sonic
events generated by RAF or NATO aircraft for [that] evening."[32] But as the

47

Flying Regulations cited by Mr Spellar made it clear, there would only be a record if the pilot had reported a breach of regulations. And, of course, if the aircraft concerned was a secret one, how could such a record be independently checked?

More details about this training exercise were revealed one week later when Mrs Jackson put questions to Mr Spellar. Asked where exactly this exercise took place, on April 7 he replied: "It is not possible, twelve months after the date in question, to state precisely where military activity was being carried out. Records kept show only that aircraft were booked to carry out low flying over the Peak District between 2030 and 2107 hours local time. No low flying is permitted over the Sheffield urban area, or any other major conurbation. Records of flying at medium level – between 2,000 and 24,000 feet – are not maintained, so it is possible there were aircraft in the area at medium level [too]."[33]

Afterwards, Mrs Jackson gave her opinion that the MoD were not being entirely straightforward in their answers to her questions. She said: "I tabled a number of written parliamentary questions about what did happen that night. The responses came from the RAF a bit reluctantly and slowly. They did admit that there were military aircraft flying over South Yorkshire that night, but they did not admit to the possibility of any of them breaking the sound barrier."[34]

Following parliamentary questions, the pressure was continued upon the Ministry of Defence with a series of direct questions to the RAF Press Office spokesmen Alan Patterson and Flight Lieutenant Tom Rounds. In October, the MoD provided a list of the locations from which complaints had been received. These included locations in Dorset, Cornwall, Wiltshire, Wales, central Scotland, Cumbria and the Midlands, but none from the Peak District or Sheffield. This information only added to the impression that a major RAF/NATO exercise was underway that evening, involving a number of military aircraft from bases scattered across the British Isles.

Flt Lieut Rounds admitted he was unable to specify which squadrons were involved, but said: "These complaints form a normal pattern. Most of the low flying is done in the Southwest, Wales and Scotland. We rarely do so, but we can fly inland between Manchester and Leeds. It's very high ground, up to 2,000 feet in places, and is a very risky place for us to be flying at low-level, especially if there is low cloud, as this could lead us to penetrate built-up areas. An additional hazard for military flying is that above this area is airline traffic above Manchester and Sheffield. However, if the weather is good this would not stop us."[35] Jenny Randles, who lives in the Peak District just south of the 'aircrash zone', confirms that low-level fast-moving military planes cross this part of Derbyshire at least once a month.

Although Flt Lieut Rounds was unable to specify the squadron involved, other sources claimed that some of the aircraft came from RAF Marham in Norfolk. They said at least two Tornado bombers took part in the exercise, which may have involved other military aircraft, including Jaguars, at various stages. The sources maintained that Marham's Tornadoes were reconnaissance and bomber aircraft, and would not be the type that would be scrambled to pursue reports of a UFO intruding into UK airspace. "They were all on routine low-flying through the Peak District and all returned safely," maintained the official line. These revelations were followed up by a direct challenge to the RAF Press Office over claims that there had been a cover-up over the Howden Moor incident.

The late Ralph Noyes, who was the former head of the MoD department at Whitehall which dealt with UFOs in the 1960s, was convinced that the UK government did know more about UFOs than they would care to admit to and, for instance, have gun-camera film, which he had seen, of unexplained light phenomena. But Mr Noyes said on the record that officials would never directly lie in response to direct questions from the Press.

Following this advice, I asked if the MoD would confirm or deny that military jets had been scrambled to pursue an Unidentified Flying Object above the Peak District, or whether one of the jets taking part in the exercise had inadvertently broken the sound barrier while exercising or pursuing an unidentified target. The full text of the MoD's reply, which is in effect a summary of their official stance on the Howden Moor incident, was:

"There has been no cover-up over this incident. We did not scramble aircraft that night to intercept a UFO. All missions were regular pre-booked training flights. We have to fly low at night to train pilots for action in places like the Gulf, but we don't fly over the urban conurbations and would have avoided Sheffield. We don't break the sound barrier over land, and we don't fly below 250 feet, although operational low flying is allowed in the Scottish Highlands as low as 100 feet.

"Our pilots know very well they should not fly at twice the speed of sound over land and would face disciplinary action or even a police investigation if this was proved. We responded to a request from the police to help them search for a crashed aircraft and we sent a helicopter along to help. We have not been chasing UFOs. We would send fighters to intercept if we picked anything up on radar screens. It has happened before, not so much now but certainly in the past. Radar is constantly looking out and can spot incoming objects, and we have a requirement to defend the UK from attack, whether that be from Libya or from Mars. We don't know what caused the sightings and the sonic booms, and the whole thing remains a mystery to us."[36]

This unequivocal reply leaves us with many questions unanswered, but supplies many clues which point towards a mundane explanation for the Howden Moor incident. It seems obvious that a concatenation of unconnected events conspired on this particular night to produce a scenario which led the emergency services to conclude an aircraft had crashed. These included a covert low-flying exercise by the military which – despite denials – was apparently associated with the mysterious sonic booms, and the presence of at least one as yet unidentified light aircraft. Mundane explanations are seldom considered an option for those who are determined to find evidence of alien encounters and government cover-ups, but in this incident there is little, if any, corroboration to suggest UFOs were ever involved.

Remarkably, on 1 October 1999, Gaynor South at Secretariat (Air Staff) 2A of the MoD – their UFO division – added comments on the sonic boom that need consideration. She noted: "You are correct in stating that military aircraft may only fly at supersonic speeds over the sea and to fly supersonic over land is in breach of flying regulations . . . You will be reassured to learn that the RAF police did investigate the incident. Their investigation found that no RAF or NATO aircraft were operating in the area at the time and that civil aircraft in the area were travelling too slow to have generated a sonic boom."

Despite the evidence of more than 200 personnel, the majority of whom were civilians, in the search for the supposed 'crashed aircraft,' of which no trace was ever discovered, there are those who continue to believe that something did crash. To the believers, the rescue teams had to be searching the incorrect part of the moors, or were deliberately sent to the wrong place while the crack UFO retrieval squads worked to remove the evidence for ET contact. The complete lack of evidence for these unlikely scenarios will not prevent them from spreading through the growing UFO rumour-mill, at every stage becoming increasingly distorted and removed from reality. The quotation from an Internet discussion of this case which opened this chapter demonstrates how 'facts' can be distorted to present a completely different picture, and illustrates how UFO myths are created and perpetuated.

Few UFO investigators are dedicated or open-minded enough to pursue a case to the extent necessary to reach the truth, which is often hidden deep within the layers of misinterpretation and distortion which make up the UFO mystery. The will to believe in the face of incontrovertible evidence to the contrary is far often likely to take precedence in the case of those believers in extraterrestrial UFOs, who have a vested interest in perpetuating rather than solving mysteries. In the Howden Moor incident, those who came to 'investigate' the case found precisely what they expected to find. Believers in ghost-planes added another baffling case to add to their files, while those looking for evidence of aliens discovered all the elements required to create another piece for the jigsaw which is the crashed UFO myth. Meanwhile, those who actually took part in the events of March 24, 1997, could only sit back and watch in amazement as a myth was created in front of their eyes.

3

A PERFECT CASE?

JENNY RANDLES

"(This case) remains today one of the most important pieces of film available to UFO researchers."   – John Spencer, for the British UFO Research Association, *The UFO Encyclopedia*, Headline, 1991.

Is there such a thing as a perfect case, or are all encounters flawed in some way? One event had just about everything going for it that a properly sceptical scientist could want. It occurred in January 1973 on the Oxfordshire/Buckinghamshire border and has been regarded by enthusiasts as invaluable proof. It featured on TV. There have been unique levels of research interest displayed. To this day its primary witness is still sure that he captured perhaps the best evidence on record of UFO reality. But did he?

A big news day ...

On the cold, overcast and slightly misty morning of January 11, 1973, the offices of the *Thame Gazette* faced two big stories. As a small market town south of Oxford, now expanding as the stockbroker belt eats up the countryside, it has rarely been home to much great news. So these two events were to grab undue attention at this rural paper.

The first concerned an aircraft from the US Air Force base at Upper Heyford to the north of town. An F-111 was on a routine training flight and had crashed in flames 25 miles east north-east of Thame. F-111s were then proving all too troublesome. This was by no means the first such accident. The danger posed to those living near a military base was slight but real. However, the impact site that day was in a farmer's field near North Crawley, now part of sprawling Milton Keynes, but then just a country backwater. Nobody was injured in the crash at 9.46 am. But the fuselage section of the plane was on fire as it fell from the sky, narrowly missing a farm building.

The other story was of rather more immediate interest to the *Gazette*. This was because it occurred well on their patch just outside town. Many youngsters at the Chilton Church of England School near Long Crendon had seen a UFO. The orange and silvery dome silently crossed their path at 9.00 am just as the bell to signal lessons was about to ring. The headmaster was convinced by the stunned looks on faces that greeted him in the playground. When a teacher arrived to say that she had seen a UFO whilst driving to work, there was no doubt. The newspaper was called to report the story.

If there was any dispute at the *Thame Gazette* over which item to pursue, that probably ended at around 11.30 am when a man called into their offices and announced that he had just taken movie film of a UFO. Peter Day was quite surprised to discover that the paper already knew about his sighting some two hours before. As soon as he reported what he had filmed from Cuddington, three miles south-east of Chilton, it was clear to the *Gazette* that all these people had observed the same thing. Now there was the thrilling prospect of moving images to back up the story. Day was assured that the paper would feature the affair and he hurried off to make arrangements for speedy processing of his film.

Apathy

On January 16, the encounter did make the local Press, but oddly went no further than that. This was not for want of trying. Peter Day had on his return home rung around various TV stations to try to attract some interest. They were disinclined because the footage had been secured with an 8mm home ciné camera and this was not readily suitable for transmission. A little frustrated, Peter gave up, ready instead to field the barrage of questions from interested UFOlogists which would soon come his way.

He was not to be disappointed by the national media for long. His film got its TV debut in 1976 when the BBC researched a documentary on UFOs called *Out of this World*. The Cuddington images were screened, but without any background story to indicate the significance of the case. The same was true in several future documentaries, such as a BBC *Horizon* programme. It was placed in front of viewers as titillation alongside many other clips of dubious or scientifically explained phenomena. As such, nobody would be aware that there were multiple witnesses who had seen the same object from different locations, let alone near-unique scientific investigation into the nature of the UFO.

It became so common to use the film as stock footage of what a UFO might look like without any attempt to investigate its origin that one BBC station showed it and incredibly told viewers that it was taken in Bacup, Lancashire – 200 miles away! All contact with reality had disappeared through repetition.[1]

Such cavalier treatment by the media soon made Peter Day cynical of the Press. Ultimately, he became disinclined to co-operate without assurances about how seriously anyone would treat his evidence. He got very few promises. Money was never a factor. All Peter Day cared about was that the public got to know the importance of his case as he perceived it. Sadly, they hardly ever did.

School's out

Let us now backtrack and look in more detail at the stories first told by witnesses at Chilton School. These were given to the *Thame Gazette* within hours of the sighting and also to investigators from the Oxford-based UFO group Contact in March. Contact had been launched a few years earlier under the odd name of the International Sky Scouts. It was founded by prolific UFO writer Brinsley Le Poer Trench, a man destined to become the Earl of Clancarty. The group outgrew this rather curious start to become a data collection agency that still thrives in Oxfordshire. They arranged a joint investigation with BUFORA (British UFO Research Association), a welcome move that was unusual for 1973. Today, such a liaison would be almost unthinkable in the cut-throat world that much of UFOlogy has become.[2]

Roger Stanway, a young lawyer and chairman of BUFORA, took on research into Peter Day and his film. Stanway was also a fine UFO investigator and a great loss to the field in 1976 when he quit literally overnight. I had dinner with him just hours previously without any inkling of what was to come. His reasons cited were new-found, deeply Christian beliefs that were incompatible with UFO study.

Meanwhile, Rick Roebuck and Derek Mansell of Contact focused on the school witnesses, getting testimony direct from them. However, they only spoke with four or five children when more than 20 appear to have been witnesses.

Peter Day was also a source of information about the schoolchildren's stories. One girl told him, for example, that the UFO was so low it "lit up the ground with an orange glow." Another said it was like a blob of fire in the sky. Peter arranged to take his developed film to the school days after the incident and a lengthy discussion ensued about who saw what. This was before any UFOlogists reached the witnesses. All who interview witnesses – to a UFO event or crime scene – have learned that giving visual cues like this and also having a long gap before investigation can occur often prove significant. Recall of what somebody sees can subtly alter to make it more like the images on film.

Chilton School is at OS Reference SP 687115. The children in the playground had the closest view of the UFO. They were mostly aged 10 and 11 years old. Some of them gave the following accounts.[3]

Louise Driver saw a ball of light that was yellow in the centre and had a darker orange/brown edge. It appeared in the west, hovered to the north over Dorton Hill and descended and rotated in a spiral motion as it did so. It also

---

UFO CRASH AT AZTEC, NEW MEXICO

One of the first UFO crash retrieval stories was made famous in the early 1950s by newspaper columnist Frank Scully. Scully's 1950 book, *Behind The Flying Saucers*, recounted how a 30-metre-diameter crashed saucer had been found on a plateau near Aztec, New Mexico, with sixteen dead aliens inside. Scully's information came from two scientists, Silas Newton and his partner Dr Gee (real name Leo GeBauer), who claimed they had discovered the crashed UFO, one of three they found in the region, and retrieved a device from it which could locate oil deposits. They told Scully the craft was constructed of a super-light, indestructible metal, and that writing inside the control room was like "Egyptian hieroglyphics."

Is it real?

● Some details of the crashed saucer are very similar to those of the Roswell Incident, yet little was known about Roswell in 1950.

● Some UFOlogists have speculated the crash rumours were an intelligence agency ploy to divert attention from the Roswell Incident.

Is it solved?

● No one in the area remembered anything about the alleged incident.

● Newton and Dr Gee were convicted confidence tricksters.

Conclusions

This case is now little mentioned in UFO literature, but the fundamental elements of crash retrieval lore such as unbreakable materials, dead aliens and hieroglyphic-like writing are all present. At the time of the event, the Roswell case was hardly known, but Newton and GeBauer spread variations of their tale to hundreds of people in the American south-west, possibly creating or adding to other crash retrieval stories from that region. This glut of interchangeable rumours and lack of hard facts cast doubt on all crash retrieval tales from America in the 1940s and 1950s. The incident also inspired a March 1950 FBI memo about the 'discovery of alien bodies' often wrongly cited as proof of a military cover-up.

Further Reading

Saucerful of Secrets. pp. 156–159. Roberts, A. "UFOs 1947–87." Ed. Evans, H. and Spencer, J. *Fortean Times*, 1987.
UFO Encyclopaedia. pp. 267–268, Clark, J. Vol. 1, *Omnigraphics*, 1998.

---

shook from side to side. Then it moved away eastwards towards Aylesbury and had more of the appearance of a flattened dome.

Paula Fox added an interesting point that the object descended briefly, then began to rise gently before levelling off and moving away. Meanwhile, Nina Sparks was adamant that inside the orange ball there was a distinct silver shape.

Most interesting of all was Tracy Perrett, who said the object seemed to "crack open into a V shape and then slow down." Then the 'V' began to fill in and move away. The V was opaque – meaning you could partly see the sky right through it – but it was darker than the surrounding object.

None of the children reported any sound. The most consistent features in the various descriptions were that the object rotated, descended and then rose upwards, and that it was less bright towards the end of the estimated two-minute duration of the sighting.

Teacher's notes

The teacher was Elizabeth Thompson. At the time she was driving to school on a small lane from the village of Ickford about three miles west of Thame. She was heading north-east and approaching the right turn at a T-junction leading into the village of Shabbington. Her OS reference was SP 655077. She was about the same distance from the UFO's west-to-east flightpath as Peter Day was to its south, but she was five miles to his west, and about three or four miles south of the children.

Mrs Thompson's first account was given to Peter Day a week after the sighting. She told him that as she approached the crossroads, the UFO was hanging in the sky ahead, seemingly motionless. It was an orange ball with a flattened base, creating a dome-like effect. It was slowly rotating, anti-clockwise if viewed from above. She turned right and stopped the car, but her view towards the object became blocked by hedges. By the time she reached a new spot giving a clear view to her north-east, the object had vanished.

Rick Roebuck interviewed Mrs Thompson and the children seven weeks later. This was long after she, too, had seen Peter Day's film. She now described the UFO as being several miles distant and in view for only about 10 seconds. But her clear impression was that it had been stationary and rotating.

Like the children, who were adamant about the time because they knew when school started, Mrs Thompson insisted that it was 9.00 am. When she reached the school at approximately 9.10, the children were still animatedly describing to the headmaster what they had just witnessed.

Interestingly, the screening of the film convinced the children that they had seen the same object Peter Day filmed. This was despite their accounts being a lot more graphic than his, probably due to their much closer proximity. Elizabeth Thompson was less persuaded when Day visited the school. She said that what she observed was bigger, more hemispherical and clearly rotating, despite these features not being apparent on the film. Mrs Thompson was never in doubt that she was witnessing a solid object. Indeed, she had tried to find a new grain silo in local farmers' fields as she assumed one of these might have been painted orange and had fooled them all. However, she admits that the object appeared to be in the sky.

After hearing about the plane crash that morning, Mrs Thompson speculated that the object had been a parachute used by the pilot to escape. This was a reasonable theory, but easily disproven. The crew did not eject in this way and were certainly still aboard their aircraft even after the UFO sighting was first reported to the newspaper. The crew did not eject until 9.45 am, and over 20 miles to the east of the UFO witnesses, in a position impossible to see from their locations. It seemed that the air crash was not a factor.

Ultimately, Mrs Thompson conceded that all of the witnesses must have seen the same UFO, and that the differences in their perceptions were mostly due to variations in distance from the object.

The UFO in focus

At the time, Peter Day was a building surveyor who lived in the village of Moreton, one mile south of Thame. On the morning in question he was visiting the site of an old water mill at Lower Winchendon, just west of Aylesbury. Here, he would meet a client for whom he was to carry out the architectural planning work necessary to convert the mill into a luxury dwelling. His appointment was set for 9.30 am.

Peter left home at just before 9.00 am to give himself plenty of time to get to the mill and have a quick look round. By just after 9.00 am, he had passed through Thame and was on the A418 heading north-east. It was here – at OS Reference SP 707070 – that Peter was first distracted by something that caught his eye over fields to his left (in the north).

When he first looked towards it, trees blocked his view, but within seconds he reached a clearer area of open fields and was able to see a ball of light moving roughly parallel with his course, but some miles to the north. No other car on the road seemed to take any notice, but the object was low on the horizon and would not be easy to see whilst driving, especially as the undulating terrain kept causing it to vanish.

As the road was narrow and there was no recognised stopping point, Peter Day could only drive on and occasionally glance to his side. But after some two minutes, with the time now about 9.05 am, he turned off to the left on a little side road toward Lower Winchendon. This was his intended route, and not a diversion to watch the UFO, but it gave him the opportunity to stop the car for a better look.

By the time he did stop, the object was disappearing towards Aylesbury. At OS reference SP 740103, Day recalled his movie camera. This was a simple model (a Japanese-built Pacemaker 200), but he had used it frequently. Day enjoyed filming aircraft ever since his own experience flying Oxfords during the war. He had shot the high-speed Red Arrows aerobatic display team in action and, indeed, took film of them to show to the schoolchildren the following week in addition to his UFO images. This was, as he put it, "to make a bit of a show for them." The camera also proved useful to record various aspects and angles of the buildings he was to work on whilst back at home, so was normally left in the car.

By the time Day removed the camera from the glove compartment and took off its plastic cover and lens cap, the object had crossed the road ahead and was moving away past some distant clumps of trees low to the horizon. He wasted a few more seconds flicking the zoom switch and then discovering it was already on zoom setting. Day worried how much film was left on his partly used super 8mm stock. But eventually he propped against the wound-down driver's window and shot.

The UFO moved behind some trees, re-emerged, then did so again as it passed a second tree clump. It was an orange ball with slight variations in colour and brightness. Despite the open window and total lack of traffic on this country lane there was no sound from the object.

Day estimated that he filmed for 12 seconds before the object vanished, although this was to prove to be a 50 per cent under-estimate. In an interview with Roger Stanway two months later, he said the word 'vanished' is accurate as "One minute it was there, the next minute it was gone." He added in a contemporary written statement for BUFORA, "This word 'disappeared' is meant in the true sense: the object moved neither up, down, nor sideways – it was just not there any more."[4]

With the UFO gone, Peter Day remained on the spot for "a few minutes." No traffic came by. Then he realised that his client was a stickler for good time-keeping and set off for Lower Winchendon. This was only about ten minutes' drive from where he was, but he arrived at around 9.35 am and had to promise an exclusive screening of his film to placate the customer.

These timings do not totally add up. A reconstruction at the site with the witness in 1983 brought only the suggestion that he must have waited for as long as 20 minutes after the UFO departed, which Day did not think he had done. Given that the timings from the school are known to be accurate, Day must have finished filming the UFO at about 9.05 am. Either he did wait at the site for a longer time than imagined or his client arrived early, causing Day to believe he reached the old mill late. It is somewhat surprising that nobody in the UFO world spotted this and suggested that Day suffered a 15-minute time lapse, thus subjecting him to hypnotic regression in search of an alien abduction. In similar circumstances this has actually happened!

Aftermath

After spending almost two hours at the mill with his client, Peter Day set off home and stopped at the newspaper to report what he had seen. He reached home about noon to be greeted by his wife telling him that a plane had crashed 25 miles away that morning. Perhaps he had filmed its last few moments. This possibility was as exciting to the witness as filming a UFO. He knew that it would be newsworthy.

Connie Day believed that she heard the report of the crash on the 9.00 am radio news, immediately after Peter left home. This is not possible as the crash had yet to happen. It was on the 10 am news as a late news flash, with fuller reports at 11 am. But the erroneous impression that the crash had been reported at 9.00 am convinced Day that he could not possibly have filmed the aircraft because it must have crashed before he left home. Even so, to make sure of his facts, he phoned the USAF base at Upper Heyford and reported what he had seen and filmed to a US major.

The USAF officer confirmed that an F-111 had been lost, but its crew were safe. He added the news that this crash had occurred at a distant location later that morning. The fire being reported by the BBC news, Day was told, had been caused on impact. In fact, this was slightly untrue, although the key facts were correct. Whilst he offered the USAF his film for study by their accident enquiry, this proposal was rejected. This has again been seen as significant evidence that Day did not film the F-111.

The key reason cited by the major for this disinterest was Day's insistence that he saw no trace of an aircraft behind the orange ball. This is evident from the film itself. As such, Upper Heyford gave the witness every indication that what he had filmed was not connected with their aerial mishap.

Nonetheless, before making his series of fruitless calls to interest TV channels in his film, Day called the Ministry of Defence in London. He was shunted from department to department before being put on to a man at an office called S4f where they "recorded details of UFO sightings." A few basic questions were asked of the witness, such as date, time and location. These, we know, would be entered on a UFO report form, but Day was never told this. That file probably still exists to be released onto the Public Record Office in January 2004 after 30 years' enforced secrecy by the British government. Such secrecy appears absurd given that the MoD had requested of Day that he post to them his undeveloped film by ordinary mail! Understandably, he declined, suggesting instead that he drive to London and screen it for them to see. The hesitant clerical officer in Whitehall said, "I'll have to get back to you on that." He never did.

Securing the film

Concentrating now on his film, Peter Day decided not to trust it to the post. By chance, the main Kodak processing plant at Hemel Hempstead was only an hour's drive away, so he chose to take the next day off work and deliver it himself. An explanatory phone call to Kodak brought a promise of co-operation from an intrigued staff member. They would develop the film while he waited so that all could see exactly what had been secured.

The footage ran for 23 seconds and shows pretty much what the witnesses described – an orange flattened ball that drifts slowly along the horizon, passes behind clumps of trees and pulses in brilliance.

There were, however, some surprises. The foreground looked much darker than Peter Day recalled. This was partly an effect of the relatively poor camera optics plus the fact that the sun was not long risen on this mid-winter, overcast day. Indeed, so dark was the image on screen that it was hard to make out any detail of the fields between the witness and the UFO.

Peter did not remember any pulsation effect such as was visible here. Possibly this is what was described as rotation by witnesses closer to the object. But there is no trace on the film of any silvery mass, a notable dome shape or of the effects described by some of the schoolchildren, such as the V shape.

However, much the most curious feature of the film was the very last frame. Peter Day had switched the camera off when the object vanished. But there is no film at all that shows the sky without the object in it. For even an experienced camera operator to be able to react so fast as to switch off within a fraction of a second between one frame and the next is almost impossible. But that seemed to happen here.

Possibly, the solution lies in the way in which the film appears to become blurred as the UFO disappears. The hedge in the foreground seems in focus as the background "smears." Peter Day likened this to a painting that has been smudged. Much debate followed in UFO circles since the film was first aired, but it is probable that this effect – however caused – merged with the dark background to make the sky apparently invisible.

After much discussion with BUFORA a theory was suggested. The originator of this appears to have been Dr J. Allen Hynek, the famous American astronomer who was consultant to the US government UFO investigation team from 1948 to 1969. He later formed the Center for UFO Studies, now one of the top groups in the world. Hynek stayed briefly with Roger Stanway during 1975.

After seeing the Day film, he commented that the blurring might be connected with a 'force field' emitted by the UFO when it departed at fantastic speed. If it moved so fast that its disappearance was not captured by successive frames of film, this effect could have bent light rays close to the UFO. Imaginative as this proposition is, there is no substance to it. Whilst Peter Day was impressed by the thoughts of such an eminent scientist and remains at least open to the possibility we have established the facts, the blurring is simply normal camera shake caused by pressing the off button. The lens pans downward as a result into the dark foreground.

Seeking the truth ...

The BUFORA and Contact investigations concluded in summer 1973. Little work seems to have been done to try to find answers. The case was considered important because, for once, we had three sets of witnesses at different locations and movie film that supported their testimony. There was no reasonable likelihood that this case was a hoax. Nor did the object on film resemble anything obvious. In 1970s' UFOlogy one collected evidence to prove a UFO. Today, the proper philosophy is a reversal of that idea. A case should be assumed to have an explanation because 95 per cent of them do. Every effort ought then be made to find the solution and its status as a UFO only accepted if there seems no reasonable alternative.

In 1977, my colleague Peter Warrington and I spent some time working with the BBC on a television documentary. We were also researching a book, published in 1979.[5] From the BBC, we discovered the existence of the Day film. Although I had been a member of BUFORA since 1969, I was unaware of the case. In fact, it was to be another decade until – as BUFORA's then Director of Investigations – I published a full report in the BUFORA booklet *Fire in the Sky*.[6]

Peter Day proved very co-operative, especially as we had formed a good liaison with Peter Southerst at the Kodak laboratories. Southerst was Public Relations Officer and had a personal interest in UFOs. He was helping us with detailed appraisals of various photographic cases. Peter Day loaned the original film for in-depth analysis.

In March 1978, Peter Warrington and I drove to Hemel Hempstead and met with Day and Southerst at the Kodak laboratories. The hours spent in analysis were helpful in clarifying a few things. There was no indication of tampering. The film was genuine. The darkness was caused by the camera and lighting conditions, but repeated recopying had altered the density as well. Tests on the final frames also established that the foreground was blurred and not just the sky. There was no need to invoke mysterious force fields.

Using figures from the site itself coupled with Kodak measurements a mathematical formula was used to fit the testimony of all witnesses together. The trees that the UFO appears to go behind are not at uniform distance from the camera. They were later measured as 0.2 miles (the nearest), with the farthest just over 1 mile. This means that the object was at least that far distant from Peter Day.

We had two weather records, the best of which was logged at almost exactly the right time (9.00 am) at Abingdon, 12 miles from Day's location. This showed near overcast conditions at 1,800 feet, slight mist with a six-mile visibility, no electrical storms, cold (3.7 degrees C), a slight breeze (7 knots easterly), and conditions that favoured thermal layers.

Combining these facts, we can come up with reasonable estimates for the dimension of the object. This places it at 3.8 miles north of Peter Day at just about the maximum visible height of 1,800 feet, with a diameter of around 60 feet (unexpectedly large), and moving at a speed of about 140 mph.

From news like this, possible solutions seemed limited, but one did occur to Peter Warrington and me. We felt this might be what we had termed in our book a UAP, an Unidentified Atmospheric Phenomenon. In other words, some kind of natural anomaly akin to what is called ball lightning.[7]

Hunting ball lightning

Ball lightning (or BL) is a rare phenomenon that occurs during thunderstorms or, not infrequently, even in their absence. It can form in the open or inside rooms. It has even appeared inside an aircraft in flight as a drifting ball moving along the aisle! Indeed, it was this case, during which one passenger chanced to be a physicist, that allowed the phenomenon to be adopted by mainstream science. Previously – right up to the 1960s – BL was dismissed by many scientists as freely as they rejected UFOs.[8]

Although there are now many well-attested cases and laboratory experiments into BL, there remain very few reliable photographs and no moving images. The Peter Day case was a tantalising possibility. For that reason considerable effort was expended to research the option. This meant by-passing the normal route of UFO investigation and appealing directly to science.

Peter Warrington and I worked in conjunction with Kodak and had help from astronomer Ian Ridpath, a UFO sceptic. The outcome was an in camera one-day seminar that Kodak set up at their London offices. We agreed to present the case in a straight, scientific manner. We also agreed not to use the event to gain publicity. In return, six or seven of the country's leading atmospheric physicists and experts in BL agreed to attend. They came from various universities and government research laboratories, such as Harwell. A representative from the MoD agreed to attend. Peter Day was on holiday on the date arranged in September 1978, but provided the original footage so that I could present the case for him. Sadly, I doubt there has been another conference of this type in British UFO history, but it is clearly a sensible way to progress such evidence.

After Peter Warrington and I screened the film several times and described the facts of the case, the matter was turned over to discussion with the scientists. There was widespread agreement. Our suggestion that BL may be misreported as a UFO was accepted as likely. It was agreed that better liaison between science and UFOlogy would be useful to ensure that physicists do not miss out on valuable data about BL or possible UAP. Sadly, no such liaison has been put in place, and science and UFOlogy still seem to exist on opposite sides of a great divide where too little mutual trust is displayed.

At the London seminar, various BL experts expressed disappointment that the film probably did not show this phenomenon. Whilst parameters remained uncertain, they believed that the ball of fire had formed in the wrong meteorological conditions. When we went to great lengths to avoid the term, they commented: "Do not be afraid of calling it an 'unidentified flying object.' That is what it is."

The seminar ended over lunch. One by one, I met with the scientists and – with Peter Day's blessing – offered them the footage to take back to their laboratories for study. Despite their brave words that we should not be afraid to call the object a UFO and none having any explanation to offer for what the film displayed, they all declined this invitation. As one man put it to me: "Next year I have to apply for a new research grant. If my colleagues knew that I was researching UFOs, then I might not get it." Therein lies the dilemma faced by UFOlogy when trying to woo support from science: regardless of how carefully we conduct our investigations, other factors can get in the way.

Chopper squad

The man from the ministry at the London seminar was Staff Sergeant Ron Stafford. Back in 1973, Rick Roebuck had fired off the usual letter about the case that most investigators send to the MoD for help when they first take on a case. On this occasion, S4f offered a "suggestion," admitting it was just that. They claimed to have checked radar records and there were no "unauthorised" military flights in the area at the time. But to them the incident sounded "consistent with that of an aircraft flying with the use of its afterburners which were switched off as the observer watched it." In retrospect, the word "unauthorised" is of note. Does it imply that there was an "authorised" flight they preferred not to discuss?

Afterburners are blasted jets of fuel that emerge from the rear of a plane and allow it suddenly to accelerate whilst in mid-flight. It was raised at the symposium, but Stafford, a munitions expert, was not impressed. He particularly noted the sideways motion. This comment was also offered by RAF air crew to whom I presented the case in May 1989 when lecturing at RAF Shawbury. One added that jets with afterburners are extremely noisy. You could not miss hearing them at such a low height.

Sergeant Stafford had the only positive suggestion to emerge from the London seminar regarding a solution to the case. He argued that there had been an experiment using helicopters specially fitted with orange searchlights. If one of these had been operating in Oxfordshire in January 1973, it could be the answer. We checked into that. Such an exercise was tried by the RAF, but quickly abandoned as unsuccessful. It did not get underway until several months after January 1973.

The new study

Soon after becoming BUFORA Director of Investigations in November 1981, I persuaded the organisation that this was precisely the sort of case upon which we should focus our attention. It offered that prospect so elusive in UFO circles of tangible evidence coupled to credible multiple witnesses.

In August 1983, BUFORA planned to stage a conference in High Wycombe, only a few miles from the site of the Day film. It was too good an opportunity to miss and I proposed two points – firstly, that a reinvestigation commence, and, secondly, that we stage a debate on the evidence at the conference. Unfortunately, the first objective was made difficult because the scattered case notes compiled by Roger Stanway ten years earlier seemed to have disappeared and it took a very long time for BUFORA to bring them back together. However, a multi-faceted research project was set in motion by our investigation team.

Sceptic and UFOlogist Steuart Campbell had been researching BL for some years and decided to explore this connection more fully. He was ultimately to agree with the findings of the 1978 seminar: that the film does not depict ball lightning. After this discovery, he quit BUFORA, but not the case.

Ken Phillips, who lived in Milton Keynes, conducted a local investigation of the F-111 crash, and I made enquiries into this incident on a broader level. These checks provided a wealth of information.

Peter Warrington set in motion attempts to have the film computer enhanced. He found a facility that would do this at RAF Farnborough. However, the cost (£1,000) was in excess of any sum that BUFORA could contemplate, given UFOlogy's meagre resources.

The main task befell a local UFO team based in Swindon (SCUFORI), who were noted for their cautious investigations. They agreed to re-interview as many witnesses as could be traced.

Allen Hynek also returned to the UK for the 1983 debate, and with John Timmerman of CUFOS went to the site with Peter Day to secure exact bearings and other readings using the better equipment that was now available. In this way, we were trying to cover as many of the bases as possible.

The SCUFORI team (Marty Moffatt, Martin Ship, Mike Williams and Charles Affleck) did a fine job collating the notes and tidying up loose ends. Their 108-page summary was sent to me in June 1984. They found Peter Day "respectable, intelligent and sincere," but noted how he had an interest in UFOs, and seemed keen to have his evidence given the status he felt it deserved.

Although SCUFORI were able to obtain useful new data from Peter Day, they were unsuccessful in getting any help from the school. By now, of course, all the witnesses had long since grown up and left. It seemed a forlorn task trying to find them. In any event, the Swindon researchers had settled on a theory about the case. They argued that Day must have got his timings wrong and seen the F-111 about to crash.

The aircrash theory

Much as I had great respect for the work SCUFORI did on the case, this idea was not one I could readily accept. It made much sense. As they rightly pointed

---

THE SPITSBERGEN UFO CRASH

During June 1952, several German newspapers reported a crashed UFO on the remote Arctic island of Spitsbergen. The craft had been discovered during a military exercise involving Norwegian jet fighters, when both their radar systems and those of their base at Narvik had been heavily distorted as they crossed the Hinlopen Straits.

The large, 45-metre-diameter disc was found to contain a radioactive source pulsating at 934 Hertz, and was constructed of an unknown and lightweight metal. No bodies were discovered on board, but unusual symbols, not unlike those of the Russian Cyrillic alphabet, were seen on the craft.

Initially, the Norwegians thought they had captured a Russian flying disc based on captured Nazi technology. This theory was later demolished when one

Colonel Darnhyl stated emphatically that the craft had "not been built by any country on earth." Other newspaper reports in years after the event related that bodies had been found in the craft, and that it was powered by a circular jet-propulsion system.

Is it real?

● The incident is mentioned in several US intelligence documents.

● Leading newspaper correspondent Dorothy Kilgallen claimed she had been told a similar story. She later died in 'mysterious circumstances.'

Is it solved?

● There are no witnesses to the event.

● USAF intelligence officers contacted the Norwegian military, who denied any knowledge of the event.

● At least one of the newspaper sources for the story, the *Stuttgarter Tageblatt*, never existed.

Conclusion

The Spitsbergen case cannot be *proven* to be a hoax. But in view of the many differing versions of the story and complete lack of *any* first-hand witness testimony, it appears to be just one of many crashed saucer stories in circulation during the early 1950s. In Nick Redfern's book, *Cosmic Crashes*, he writes of finding an American National Security Agency document with the word 'PLANT' written in the margins. Redfern suggests the Spitsbergen story may have been disinformation to "cloud the rumours surrounding crashed-UFO incidents."

Further reading

The Spitsbergen Saucer Crash. Moore, W. Moore Pubs., 1986.
Cosmic Crashes. pp. 294–298. Redfern, N. Simon & Schuster, 1999.
"The 1952 Spitsbergen UFO Crash." Braenne, O. *UFO Brigantia*, Nos 51 and 52, 1992.

---

out, the probability of two unusual events occurring so close together in time and space was remote. Consequently, it was logical to assume that they were connected and, in fact, a single incident. From this logic it followed that the UFO *was* the crashing F-111. The main problem with this idea was that Day was certain of his timings to within five minutes, as were the schoolchildren. There was no dispute when and where the plane crashed – at least 40 minutes later and 20 miles away. For the film to depict the F-111 so badly on fire it must have been taken moments before it crashed. No furiously blazing aircraft could have remained airborne for another 40 minutes. This theory – rational as it was – just did not slot together.

Ken Phillips and I were having success tracking down details of the F-111 accident, even though USAF Upper Heyford were not unduly helpful. Captain Paula Hoffman, Chief of Public Affairs, had confirmed the accident, but would not supply any records. She directed the investigation to Norton Air Force Base in California, and advised that a considerable fee would be charged even to search and possibly find nothing. She added that "You will only be given information consistent with Air Force and Department of Defense policies." In other words, what they chose to release. It seemed this was a dead-end, especially as no accident report was available by normal routes, even though the crash had been on British soil.

Nevertheless, an attempt was made. Norton supplied SCUFORI with a brief note indicating basic details. This gave the aircraft type (F-111E), its precise impact point, heading and crash time. It added: "The aircraft was burning in the mid-fuselage area when it emerged from an overcast at approximately 2,000 feet in a steep nose-down attitude. The ejection capsule landed upright in the back-yard of a home in the village of North Crawley. The crew was not injured."[9]

Ken Phillips and I pursued sources that included locals in the crash area, newspaper archives and aviation journalists who had been interested in the spate of accidents to F-111s. It was these sources that revealed why the USAF were being so circumspect about releasing the accident report, even through the US Freedom of Information Act, which should have made it publicly available by 1984. The crew had ejected to safety using a then revolutionary system developed from the Apollo space programme. Rather than jump using parachutes, a rocket-propelled module was thrown clear of the damaged jet and guided safely to earth. The secrecy over this developing technology was an issue.

Indeed, we discovered some interesting points about the flight. Two aircraft were involved, a second F-111 being in close companion to the first throughout.

In fact, eyewitness Paul Hunt, at Bletchley, reported how "just after 9 am" he saw "two F-111s streaking low over the town." One of these had an engine that looked "white hot" and the other was flying below, seemingly looking up and inspecting the damage. For the next 40 minutes these two planes circled the area, mostly above cloud. At about 9.45 am, there was a loud explosion over North Crawley. Farmer Gordon Adderson claimed "then there was a sudden roar and out through the low cloud came the plane, all on fire. It dropped in a mass of flames."[10]

Such were the concerns of locals about the low level at which these F-111s flew above their communities that the MP for Newport Pagnell wrote to the MoD to obtain assurances that such flying was not permitted. The MoD not only gave confirmation of this rule, but added that the two F-111s circling the skies on this date did not fly too low, despite what eye-witnesses claim they saw.

Brigadier General Darrall S. Cramer – Vice-Commander of the 17th Air Force – was in charge of the accident investigation. Concerning its cause, the only solution that he publicly cited was a rudder defect that left the plane flying in a tight circle, slowly using up fuel in an effort to make an emergency landing.

These facts established that all of the main drama happened well to the east of the UFO sightings and 45 minutes later. But there was another possibility. If the UFO had been in the sky at the same time as the stricken plane, might it not have caused the initial malfunction? UAP – including BL – have been known to emit electrical energy fields. Could they scramble the complex electronics on board modern aircraft? The big clue here was a comment by Darrall S. Cramer that he was seeking to find out why his plane developed a fault 45 minutes *before* it crashed. This news placed the starting point of the ultimately catastrophic incident at the very same time as the UFO was being filmed. Coincidences were beginning to pile up.

But was this a coincidence? SCUFORI rejected the idea that the UAP triggered the plane crash because nobody saw both the UFO and the aircraft together. This is very reasonable, but the MoD had insisted that the F-111 was *not* flying low and the UFO was filmed just at the cloud ceiling of 1,800 feet. If the aircraft was well above that level, would it have been visible? Perhaps not. But it should have been heard because F-111s make a great deal of noise. Yet no witnesses described hearing any sound behind the UFO. Peter Day was some miles to the south. The children were not. They were almost underneath the object. Why did they hear nothing?

Back to the children

The only way to resolve these difficulties was to try to track down the children. Fifteen years after the sighting this proved possible when I was able to work with the BBC in London using their resources.[11]

It is estimated there were 24 children. As of 1988, these were all in their mid-twenties. Some had married and moved away. As many were female, they had also changed their names after marrying, making it harder still to find them. But I spoke to seven. This was invaluable.

Headmaster Edwin Bennett confirmed that he was impressed by the immediately consistent stories of the children and his teacher. But he saw nothing. The UFO had "dropped behind a valley" just before he got into the yard to assess the cause of the excitement.

Paula Fox told me that she saw a "bright light, travelling and spinning at a very fast rate." She added that it was far more dramatic than visible on the Day film, saying: "It was not really orange at all. It was definitely silvery white ... In no way was it just a ball of fire. I would say it had a saucer shape with a dome on it."

This news certainly explains why Paula still believes she saw a UFO. It is also apparent that if we did not have the Day film for comparison or Paula's more modest statements made at the time of the sighting, we would now be forced to evaluate her current story describing a classic saucer-shaped UFO. How often does this occur when UFOlogists chase up sightings that are weeks, months or even years old?

I asked some of the witnesses whether the excitement of the occasion had led to much noise and hubbub in the school yard that would have drowned out any sound from the UFO. But they commented on how an air of hushed silence descended on the crowd. Everyone just stared at the sky.

More detailed stories advised the following: that the most stunning difference between the sightings and the film is how much bigger the UFO was as seen from the school than from where Peter Day had filmed it. The object moved under the low cloud, dipping at one point as if about to land, but still swirling. Then it crossed a patch of sky without cloud, a small hole in the overcast. Here it "just vanished instantly." Paula Fox said she was so scared, "Frankly, I expected it to land and Martians to get out."

Nina Sparks insisted she is sure the witnesses would have heard a loud noise. But there was nothing. She added that the UFO "was just like a great big orange blob," and that it "travelled across the fields in a curved path. It seemed to be descending as it moved." Nina had no explanation to offer.

However, eventually I did find one witness who claimed there was a noise. This girl, Tracy, reported that the big ball of fire she saw "made a very big roaring noise." This would certainly be consistent with an F-111, but it begs the question why she is the only one of the seven I talked with who heard anything and why, according to the other witnesses, everyone else had felt that the UFO was silent.

I ended up with no doubt that all of these witnesses are sincere, but inevitably passing years and changing memory have played a part. In the testimony I obtained 15 years on, there were serious disagreements. Most thought the UFO was just an orange ball. Two reported a definite silvery shape. Most thought the yard was quiet, but one said it was noisy. All bar one said the UFO was silent, but this witness claimed it made an extremely loud roaring noise that it seems hard to imagine nobody else would have heard. Trying further to sort out this confusion is probably doomed after so much time.

Wild ideas

Perhaps the oddest theory to explain what was seen on that January day was provided by Adrian Berry, Science Correspondent to the *Daily Telegraph*. After seeing the film in 1987, he announced that "it was a meteorite. I think it crashed into the earth."[12] Berry has little time for UFOs, and attacked the subject on more than one occasion. It is also true that bright meteors, or bolides, have triggered UFO sightings as evidenced by other cases in this book. But here that idea flies in the face of all the facts, which I expect Adrian Berry did not have at his disposal when jumping to this conclusion.

For a start, a three-minute duration would be long for a fireball meteor. If this were the case, thousands would surely have seen the spectacle, not just a few in one narrow part of England. This is not to mention the awkward fact that the 1,800-feet ceiling and near complete cloud cover makes a bolide impossible. It could not have been below that height or disaster would have occurred when the fireball inevitably struck the ground. If it had been miles high in the atmosphere, as most meteor sightings are, burning up before any impact, it could not have been visible through thick cloud. We can safely reject any possibility of a meteor.

Even more worrying are those who cling to the ball lightning theory. John Spencer, who has been a senior officer of BUFORA for many years, should be aware that this option was explored in great depth by his own group. Almost unique to this case, we can argue from a position of real strength that the cause was definitely not BL. But in his *UFO Casebook*, Mr Spencer devotes space to the theory and even has an artist's reconstruction of it. Nick Pope, MoD-man-turned-UFO-writer, also wrote an account of the case suggesting the ball light-ning solution and was surprised when I pointed out the amount of effort expended to reject the prospect. Hopefully this chapter will make UFOlogists aware of the true facts of this case.[13]

The clue

Just over a year before the Peter Day film was taken, another case took place that may well provide the key. It was also investigated by BUFORA at the time, and I published a detailed case report on the findings when I accessed the case files and realised their relevance. They certainly gave me the clue that I needed to try to unravel the Day film.

On October 26, 1971, a major flap of UFO sightings was occurring in the Midlands. A TV debate was even staged from Banbury to discuss the sheer number of reports. This background probably explains why on that sunny day, around noon hundreds of witnesses jammed police and media sources to report a UFO. They came from Rugby in the north to Wolverton in the south. I was able to plot these consistent reports onto a map from the BUFORA files. A course for the UFO was easily detected, moving north of Banbury towards Northamptonshire. Given calculations from so many different witness loca-tions, the object was at 25,000 feet. It also changed direction at least twice, heading north, then back south, before resuming its course easterly. This explains why some observers insisted that the UFO hovered. It may have *appeared* to do so from their vantage point as it moved away or towards them in line of sight.

Visually, the UFO was described as like a silver or orange ball sprouting flame and accelerating like a rocket. But the real fortune came because at the time of this event an ATV camera crew were in a field at Enstone making a farming documentary. Camera operator Noel Smart secured excellent footage of the UFO, including some sequences showing the sky and the object with all the foreground included.[14]

This film has frequently been seen in TV programmes (again without the background), and I have debated it with the camera crew. I am satisfied that this case has been solved, even though many of the witnesses, including the ATV crew, are not persuaded by my arguments. The film shows the 'rocket trail' stopping and starting, a fact probably explained because I discovered that the weather conditions for the day indicated that vapour trails could only form above 26,000 feet. This object was flying just on that limit, making any trail that was produced an intermittent one.

Moreover, the MoD confirmed that an aircraft was dumping fuel at that time. It was releasing this in an emergency operation by venting it from the rear and igniting it with the afterburner. Not only did this cause the jet to speed up, as seen on camera, but the fuel burnt off and evaporated long before it reached the ground, which is vital in such procedures. Dumping burning fuel onto civilians is not recommended policy, hence the need to do this dangerous tactic at great height. The aircraft in 1971 was an F-111 from Upper Heyford.

Behind the Day film

Could this have a bearing on the January 1973 incident? There are problems. The very low height of the phenomenon seen around Thame is crucial, not to mention the lateral motion of the orange/silver dome and three-minute duration. The witness testimony and film have little resemblance to the ATV images.

Nevertheless, the connection between the starting point of the problems experienced by the F-111 and the UFO sighting must be considered. When compiling the BUFORA case report into the Day film, I had to take seriously the idea that the stricken aircraft dumped fuel and that this somehow appears on the film. I made it my final point in the case history and cited the reaction of the witnesses when I put this theory to them.

Peter Day was unimpressed. Paula Fox told me: "I suppose it could have been burning fuel ... but it did not seem like it to me. It seemed to be a solid object, not a mass of flames." Nina Sparks added, "I can accept that it might have been ... but surely I would have heard the plane?"[15]

So was this what happened – and if so how? Fortune intervened. On June 2, 1989, police throughout Oxfordshire were flooded with calls from people who saw a ball of orange fire that lasted for a few seconds. This was investigated and then explained by the MoD as "an American fighter pilot from RAF Upper Heyford who was igniting jettisoned fuel with his afterburners." This new case even further enhanced the likelihood of a solution to the impressive Day film, but without evidence from the F-111 crash report we were left guessing. The breakthrough had yet to come.

The answer

In 1991, the F-111 report was released by the Freedom of Information Act in the USA. Steuart Campbell obtained it, but did not send copies to the UFO community or alert colleagues at the BUFORA team that had so painstakingly felt its way towards the truth. Instead, he wrote an article for a photographic journal and attacked BUFORA for not having solved the case despite knowing that we had put on record our proposal of the fuel dump option.[16] Indeed, Mr Campbell left readers with the assumption that BUFORA considered the film as some sort of unexplained phenomenon when, in fact, we went to considerable lengths to beat a path towards the explanation through years of sheer hard work. All we lacked was the accident report to add the final clue. Aviation sources released it a few months later. This supplied all the information that we needed to put the final few pieces of the puzzle together.[17]

Sewn 11 and Sewn 12 were the codenames of the two F-111s that left Upper Heyford at 8.58 am. Almost immediately on takeoff, a fuel line nut on Sewn 12 malfunctioned, causing a leak. The port engine caught fire. Major Bob Kroos, who was flying Sewn 12, reported: "Have a fire on left engine … we are single engine and might be losing control." The fire was quickly put out by the on-board equipment, but it burnt through the rudder wiring, making the F-111 unsteerable. The pilot of Sewn 11 advised, "Dump fuel if you have to, Bob." The major confirmed, "We're dumping – we're gonna go ahead and dump."

The aircraft was still at a low height (2,000 feet), having just begun its climb. Dumping fuel so near the ground is illegal unless there is no alternative. Clearly Sewn 12 was suffering a dire emergency. Plotting its flightpath from Upper Heyford, the F-111 was due north of Thame at about 9.01 am. It was precisely where the UFO first appeared at this moment. From this point onward, little more is said about the fuel dumping in the accident report except that a 'flamer' emerged and the dumping was ended at 9.04 am. All of this is totally consistent with the sightings made by the witnesses in this case.

Why the dumping stopped is not reported, but the F-111 was now locked into a permanent long turn, heading east. Inevitably, it flew in circles over Newport Pagnell for the next 40 minutes with Sewn 11 in close attendance watching every move. Given the gun camera film of the entire accident taken by the second aircraft at very close quarters, it is now apparent why the major at Upper Heyford was not too interested in Peter Day's film when it was offered to them. He had much better evidence already.

The long circuit aimed to burn up fuel without ejecting it in order to make the plane safer for an emergency landing. At 9.46 am, with little fuel left, Major Kroos struggled to get the gear down to try for a landing back at Upper

Heyford, but this proved difficult, thanks to the burnt cables left by the fire. Now at 5,000 feet and so out of sight above the cloud, Kroos tried a desperate tactic to regain control, but his aircraft went into a catastrophic dive and he was forced to eject.

Does this explain the UFO sightings? By all logic it has to. The fuel dumping occurred during the precise same three-minute window at the same location as the object seen and filmed by the witnesses. This simply must be what they observed. The fact that the F-111 had lost an engine meant that it was not making as much noise as it would normally have done, although the second F-111 was presumably nearby.

How did the dumped fuel lead to the UFO sightings? This still remains contentious, but given the weather conditions, we know that Sewn 12 chanced to be at the point of the thermal layer where thick cloud began just as fuel dumping was initiated. The 'flamer' could have been temporarily trapped by the thermal layer at the cloud base and effectively created a floating mass of fire that was dragged along by the jet's slipstream, illuminating the base of the cloud. This would be a spiralling mass of orange fire that would descend, be buoyed up on air currents and then disperse. This is just what the school-children described.

It is also quite reasonable to suppose that once the danger created by dumping fuel at low height in these prevailing weather conditions was recognised by Major Kroos, he stopped doing it. He then attempted to burn off fuel more slowly without a threat to people on the ground, or indeed to his aircraft. So why was this truth not quickly admitted? Perhaps there was a combination of circumstances here that made that inevitable.

The horrific news that this blazing fuel had almost been dumped on top of a school yard packed with children was probably recognised immediately from the media stories about the UFO case. It is unlikely that either the MoD or USAF would have been very keen to trigger a public relations disaster by having to admit this fact to a Press already worried about F-111s falling out of the sky with regularity.

Even to this day, 27 years later, there have been no public revelations about just how close Chilton School may have come to disaster on that January morning. Of course, this incident was nobody's fault, but just a terrible combination of circumstances. Major Kroos acted properly in desperate circumstances and put his life at risk flying a damaged aircraft for 40 minutes rather than continue to dump fuel and endanger the public.

But perhaps it was felt better that this story remain a UFO legend than the quite terrifying reality be made clear. The way of the world forces the military to fly jets like incendiary bombs laden with fuel at fast speed and low level over several parts of the UK every day. Those flight paths include school play-grounds. From time to time emergencies force the crew to eject fuel and engage in risky operations – as in the high-level incidents of October 1971 and June 1989 – and no doubt others of which we are unaware.

The events of January 1973 were not at high level ... and a very definite threat to people on the ground. UFOs might well provide a convenient distraction that would turn attention away from the tragedy that might have been. In that sense, in seeking to prove an alien reality or to defend the unexplained nature of cases such as this, UFOlogists may well serve as unwitting puppets to the powers-that-be. That is yet another very good reason why it is important to seek to unravel the truth behind all UFO encounters. At times that truth can be even more disconcerting than the presence of an alien star-ship.

4

SPEARHEAD FROM SPACE?

David Clarke

"UFOs possess a powerful, seductive lure that continuously changes to confirm our deepest fears or realise our greatest desires. Only the form changes to reflect the social and cultural context ... What people claim to observe and experience are reflections of popular social and cultural expectations of a particular era."   – Robert E. Bartholemew and George S. Howard.[1]

At the beginning of the 21st century, we cannot escape the influence of UFO mythology. As we grow up, we are surrounded by images of space travel, aliens and flying saucers in films, TV, video games and on the Internet. This has led to a situation where some individuals and groups now accept the myth of extra-terrestrial visitors not as just a theory or a possibility, but as an established fact of which the 'proof' is seen as overwhelming and beyond question.

Science fiction writer Isaac Asimov once remarked he had been told that so many people have seen objects which looked like spaceships "there must be something in it." His retort, however, was: "Maybe there is, but think of all the people in the history of the world who have seen ghosts and spirits and angels. It's not what you see that is suspect, but how you interpret what you see."[2]

The establishment of this level of certainty of belief in the reality of alien visitors among UFO experts, groups and the public in general has created a template into which any number of disparate aerial phenomena can be interpreted as 'alien' in origin. However much ET believers who set out to 'investigate' UFOs attempt to behave in a neutral manner, their very presence helps to lend credence to the fact that a 'flying saucer' (a term synonymous with 'spaceship') has been observed. This situation is encouraged by the tabloid Press, who distort UFO sighting reports out of all recognition, and whose primary motive is entertainment, not education. This has helped to develop a stereotype of the 'flying saucer' manned by 'little green men' so beloved of Fleet Street, and of the image of 'UFO spotters' as eccentric cultists scanning the skies with binoculars.

Determined attempts to reach the truth behind the headlines have often floundered in the murky waters of belief and speculation, but there is no substitute for basic fieldwork using the techniques of an investigative journalist. In the 1920s, Charles Fort, the great collector of the strange and unexplained, said that to measure a circle one can begin anywhere. Few attempts have been made to study waves of UFO hysteria from a viewpoint based on sociology and psychology, and fewer still have attempted to track the source of the mystery back to the original eye-witnesses. During the US great wave of 1966–7, journalist John Keel attempted to measure the circle by subscribing to Press cuttings services, but within weeks he was overwhelmed by literally thousands of reports, many of them little known outside their local communities.[3]

Beginning in August 1987, I had the opportunity to study a 'wave' of UFO sightings as they engulfed the British Isles. I collected and carefully monitored the influence of newspaper stories and watched the growth of a general atmosphere of public and media interest, which continued unabated until the spring of 1988.[4] The large number of UFO sightings reported early in 1988 (with several hundred observations in Yorkshire alone) may one day be classified by historians as a 'UFO flap.'

A more balanced view would suggest the sightings were fuelled by media publicity, which coincided with a period of clear night skies and two prominent bright planets (Venus and Jupiter in conjunction) during January and February 1988. Additionally, for several weeks prior to this wave of sightings newspapers throughout the world prominently featured a story claiming that a "UFO" attempted to hijack a family travelling by car across the Australian outback.[5] The London tabloid newspaper, the *Daily Star*, followed this with a series of lurid articles dealing with the more sensational claims made by the proponents of UFO abductions, replete with illustrations culled from the pages of science fiction fantasies. Adding to the concatenation of distorting influences was the presence of a host of low-flying aircraft and RAF exercises which, it could be argued, created the 'wave of sightings' without the influence of any real 'alien' flying objects.

The UFO flap of 1987

Newspaper cuttings show that throughout August scores of people all over the British Isles believed they had seen UFOs. For example, in August 1987 the new Skyship 200 dirigible airship belonging to Airship Industries in Bedford embarked on a night-flying exercise over Northamptonshire. This produced an outbreak of UFO reports in Northampton and Kettering which described a weird "cigar-shaped object with flashing lights flying at very low altitude." However, as a spokesman for Airship Industries told the Press soon afterwards, "The Skyship has been widely reported as being a UFO … but reports

THE SILPHO MOOR SAUCER

This alleged UFO crash, which dates from November 21, 1957, took place near North Yorkshire's infamous Fylingdales radar station.

Whilst climbing Reasty Hill at night, Frank Dickenson's car sputtered and ground to a halt. Dickenson had seen a glow in the sky, which crashed onto the moor at the same time as his engine failure, so he rushed to where the light seemed to have come down. Just 30 metres from the car, he discovered a small saucer embedded in the ground. Near it, in the pitch blackness, were a man and woman, who also claimed to have seen the crash.

Dickenson ran for his colleagues, but when they returned both saucer and couple had disappeared. Dickenson eventually tracked them down via the local Press and bought the saucer for £10. Psychologist John Dale, who later saw the object, gave its dimensions as 46 cm in diameter and 23 cm deep, and its weight as 16 kg.

The UFO was a classic saucer shape and constructed of metal, some of which was plastic coated. A metal coil found inside bore hieroglyphic symbols, inside of which were seventeen wafer-thin copper sheets also inscribed with hieroglyphics.

University analysis revealed the saucer to be composed mainly of lead, but noted that the copper was unusually pure, with no tin or nickel impurities. The symbols were translated by a linguist and revealed a 2,000-word message purporting to be from an alien named 'Ullo.' Following analysis, the Silpho Saucer vanished into obscurity (reputedly after display in a seaside fish and chip shop!) and the case has never been satisfactorily explained.

Is it real?
● The saucer existed and was handled by several UFOlogists.
● The analysis indicated it was of a high standard of workmanship.
● The hieroglyphics were a genuine unknown language that had to be translated by linguists.

Is it solved?
● The lead composition of the object precluded it having come through the earth's atmosphere.
● The saucer was not damaged following its 'crash,' suggesting a hoax.
● There was no sign of any propulsion system.
● 'Ullo' may be seen as a play on 'Hello,' also suggesting a hoax.
● In 1999, Andy Roberts was told by a source at the Scarborough Evening News that the Silpho Saucer was a hoax involving journalists from that newspaper.

Conclusions
The Silpho saucer is an excellent example of a fairly sophisticated hoax, showing the lengths people will go in creating a UFO story. Although the hoax was intricate and costly, there was no discernible motive. Like the truth, the actual Silpho UFO is still out there to be found.

Further reading
"The Silpho Moor Mystery: Ullo's Message To Earth." p. 4. Flying Saucer Review. Vol. 4, No 2. Mar/Apr 1958. "Mystery of the Silpho Saucer." pp. 15–19. Randles, J. UFO Brigantia, No. 35. Nov/Dec 1988.

of the craft flying at treetop level are wrong, as the craft never dropped below 2,000 feet."[6]

A prime example of UFO folklore in the making can be found in the events of the evening of Wednesday, December 9, 1987, when more than 300 people in the Midlands counties of Nottinghamshire and Humberside saw a spectacular formation of coloured lights "as big as a football field" in the sky. This UFO appeared to move slowly through the sky in an easterly direction, accompanied all the while by a deep humming sound. Police in Hull were inundated with phone calls describing the UFO, which one observer described as a triangular object 250 feet long, surrounded by red and white lights, making a "roaring noise." Newspaper headlines proclaimed "Mystery Sky Sight Remains a Mystery" (*Hull Daily Mail*) and "Riddle of Huge UFO" (*Daily Mirror*). Despite the massive media hype which featured unfounded statements concerning spaceships and claims of official cover-ups designed to hide facts about the sightings from the public, the truth, as so often is the case, was stranger than fiction.

RAF Binbrook in Lincolnshire told investigators from the Independent UFO Network that they "... were satisfied that the reports concerned two USAF KC-135 tanker aircraft, each with up to seven F-111 [fighter] aircraft in trail; these aircraft were from Mildenhall (Suffolk) and were operating with Midland Radar."[7] This prosaic explanation was confirmed one month later by the Ministry of Defence in London. However, in the meantime, when Jenny Randles was interviewed by Radio Nottingham in a live broadcast following in the wake of the sightings, the presenter refused to accept the solution to the mystery! The implicit assumption was that, as a UFO investigator, Jenny should be supporting the belief that the lights were "alien spaceships." Similarly, many of the witnesses in Humberside would not believe that they had seen refuelling aircraft. They were utterly convinced they saw a huge UFO. One female witness even told investigators, "It couldn't have been aircraft because I saw one join onto the other!" This illustrates the process whereby belief in UFOs and alien visitors is created, fuelled and transmitted in popular culture.

By the middle of August 1987, the London newspapers had begun to feature a tongue-in-cheek story connecting the forthcoming Harmonic Convergence (a planetary alignment due to take place on August 17) with a Hopi Indian prophecy predicting the end of the world. One Dr Jose Arguelles was quoted predicting that "a wave of UFO sightings and a great, unprecedented outpouring of extraterrestrial intelligence" would accompany our imminent demise. The world did not end with the Harmonic Convergence, but the 'wave of UFO sightings' certainly did occur right on cue. In Derby, a city in the English Midlands, late on the evening of Sunday, August 16, 1987, dozens of people reported seeing a brilliantly-lighted UFO pass overhead. In Chaddesden, there were reports of people rushing into the street screaming as the UFO – said to be "as big as a car" – hovered above them. After manoeuvring at low level, the object disappeared in a north-easterly direction towards the village of Spondon.[8]

It was shortly before 11 that same night when Audrey and Trevor Boon, leaving Derby on the A608 at Breadsall Hilltop, came face to face with this UFO. Mrs Boon told us: "... All at once I saw this thing in the sky which I thought was a plane. I said, 'That plane's getting low,' and then I said, 'Oh my God, it isn't a plane.' My husband – who was driving – then looked at it, and we then thought that it must be a UFO. It was massive and the lights were brilliant, and it hovered ... All I saw at first were these very brilliant lights either side of it and a red glow underneath. Then as it hovered and moved I saw more lights, looking like big cabins all lit up like a double-decker bus in the sky."[9]

This weird object hung in "the beautiful clear sky" as if frozen in time and space. The couple were left with the impression that its crew were watching their movements as the car passed within an apparent hundred feet of the object. "It was lit up all the way round," Mrs Boon added. "There's no plane lit up like that. If it was a plane, it would have crashed anyway, because it was too low to come into anywhere. It was going towards a built-up area, and it was so low it would have hit the cathedral."

A large unidentified flying object apparently wanders lazily across the night sky of a major city. The Civil Aviation Authority and the police (as we later discovered) were unaware of its presence or its movement, and were apparently "unconcerned" about the possible hazard posed by its meanderings at such low altitude over a densely populated city centre. What of the Ministry of Defence in London? Shouldn't they be interested?

"You mention that you are concerned that alien or unidentified craft seem able to pass undetected through the UK Air Defence Region," Clive Neville of AS2 (Air Ministry, Whitehall) reassured us in a letter of March 1988. "I must point out that the UK Air Defence Region is extremely well protected and it is highly unlikely that an intruder of conventional origin would be able to travel to the UK without being spotted. I would suggest that many alien craft could, in fact, be UK military aircraft, civil aircraft, meteorological balloons and a host of uncontentious objects ..."[10]

The mention of meteorological balloons provided a vital clue towards the explanation for these puzzling sightings, for one newspaper report mentioned the presence of "unusual atmospheric conditions" during the weekend of the sightings. Immediately, I contacted Nottingham Weather Centre, who checked their instruments and confirmed the existence of "a marked temperature discontinuity in the middle atmosphere, this being a rapid rise of temperature over a short vertical distance." Such temperature inversions, said a senior meteorologist at Nottingham, "... result in the anomalous propagation of energy such as light waves ... and this outlines our objective view into what caused the sightings."[11]

Did the temperature inversion produce a mirage of some distant light source or aircraft, which gave the witnesses the impression they had seen a large object at low altitude? But even if this was the case, it does not affect the witnesses' "experience." They "saw" neither mirage nor aeroplane: they believe they encountered a UFO, and continue to do so to this day.

Lights observed at night can be very deceptive in terms of estimated distance from the observer, speed and altitude, particularly when there is good visibility. Under unusual weather conditions, aircraft at a distance away may appear to be very close and of huge size, the strangeness being enhanced if no noise is heard. Combined with a will to believe in UFOs on the part of many observers, the ideal conditions for misperception are apparent behind many 'UFO flaps.'

The flap of 1988

A further good example of the misperception of an everyday celestial object took place after midnight on Saturday, January 2, 1988, when, according to the *London Evening Standard*, "a total of eight startled police officers watched a grey-green, pink and electric blue object as it span, banked and hovered" above Kensington in London. A 16-year-old girl mounted a telescope on the doorstep of the family home after spotting a light in the sky, and soon called in the local police as additional witnesses. A Scotland Yard spokesman said later: "Police officers saw an unidentified flying object after a call from a member of the public. We are unable to say what it was."[12] In this case, it took an investigation by UFOlogists to provide the police with the explanation they so badly wanted!

This story made headlines in most of the national newspapers the following day, and even appeared on the main BBC six o'clock television news as a bona fide UFO report. As a result, the sighting became the genesis of a further period of renewed media interest in UFOs and alien visitors, which continued unabated until the latter part of February 1988. But what was really going on? Despite the media hype, rapid investigation of the Kensington 'UFO sighting' by BUFORA investigator Mike Wootten established that what was seen was nothing more unusual than the planet Jupiter. The telescope used had been out of its box only once before and was found to have significant lens distortions. A check with BUFORA's computerised astronomy programme left no doubt that the bright evening planet was in exactly the same position in the sky as that in which the 'UFO' had appeared. The final blow came when the 'UFO' occurred again in the same place the very next clear night that came along.[13] The media, however, were not interested in providing the solution to the case. "Police see alien spaceship" was the headline that was required. "Police think they see alien spaceship, but were mistaken" does not quite have the same impact, or sell as many papers.

The UFO wave continued on Monday, February 8, 1988, when the national Press reported yet another police sighting, this time from Sheffield in South Yorkshire. "May the Force be with you," headlined the *Daily Mirror* in a story describing a sighting at Ecclesfield village, in which two police officers spotted "a V-shaped machine hovering just yards away, with a row of flashing red lights along its side which lit up the entire area."[14] Another report told of how "a few hundred yards away outside the police-station, a police constable saw the spaceship hover for about 30 seconds before flying off towards Grenoside ... Officers in the control room laughed in disbelief until they, too, heard the whirring noises that accompanied the ship."[15]

Other people across the northern suburbs of Sheffield were also watching this mystery object. One lady saw it approach from Grenoside and "continued to watch in amazement as it hovered above Wharncliffe Woods just for a second and then turned left towards Hillsborough." Losing altitude, it emerged through broken cloud above busy Wadsley Lane in Hillsborough, where Linda Underhay and Gail Fairbrother were so shaken by what they saw that they were forced to stop their car in the middle of the road. "Suddenly," said Linda, "bright lights approached over the top of the houses, and appeared to be attached to a very large diamond-shaped object. Four red lights were visible which appeared to be at either corner, one green light in the middle and white lights surrounding it … We seemed to be looking at the underside of a huge object which was visible behind the lights."[16]

Typically, the release of this story to the media resulted in a flood of calls to Ecclesfield Police Station from TV news, Fleet Street and UFO societies. The police are, like pilots and the military in general, regarded as trustworthy and qualified observers, and unlikely wrongly to identify objects in the sky or put their jobs or reputations at risk by making a UFO report which has no foundation. Wisely, South Yorkshire Police refused to comment after the police sighting was initially leaked to the local Press, but almost immediately local UFO enthusiasts were quoted to the effect that "… we are checking aircraft and satellite movements today, but from what we can gather there's no way this could have been anything else but a spaceship."[17]

However, a completely different interpretation was given by another eye-witness, Norman Athey, who with two others saw this 'UFO' from a hill commanding a fine view over Ecclesfield towards Rotherham in the east. He described to us how he "first saw it a long way off in the direction of Kimberworth Hilltop. From there, it came directly towards us, flying very low until it was directly overhead. It passed overhead, which was when I was able to see it was an aeroplane, its hazard-flashing lights giving a peculiar effect as they reflected off the fuselage and wings." Mr Athey added that "because this, at first strange-looking flying machine, proved to be an aeroplane, it was only on seeing the newspaper articles about UFO sightings, and particularly the direction of the flightpath described there, that we decided it was the same object we had seen. Its flightpath should have put it over Ecclesfield police station and probably over Wharncliffe."[18]

This low-flying aircraft was never identified, although a Flight Lieutenant at RAF Finningley admitted to me privately that the craft had been tracked on radar and "would appear to be a turboprop commuter aircraft or light transport … [flying] directly from East Midlands Airport to Leeds Bradford."[19] Both airports subsequently denied any of their scheduled flights had caused the sightings, and said low-level flights such as this were more likely to be "of military origin."[20] At the same time, the MoD were emphatically denying involvement, claiming that "the RAF and NATO do not conduct exercises over major urban areas at any time." Meanwhile, the police and other puzzled witnesses were left scratching their heads.

Despite the bland denials, and the contradictory statements of both the Civil Aviation Authority and the MoD, my field investigations during the

UFO wave of 1987–8 left me in little doubt that a large percentage of the many reports related to misidentifications of aircraft, both military and civilian. It soon became clear from the testimony of numerous witnesses in the Dearne Valley area of South Yorkshire that some of the sightings were caused by an aircraft which had performed illegal low-level manoeuvres while flying on a regular route towards Sheffield on two particular evenings every week. CAA regulations stipulate that pilots of civil aircraft could face prosecution if they fly below 1,500 feet when above built-up areas, and yet this particular plane was flying so low that witnesses feared it was about to collide with chimney pots. One fire-fighter who saw the craft said "it carried lights on its wings in a V-formation, flashing red and white on and off, and made a loud engine noise when overhead, but when seen at such low level it looked very unusual."[21]

Jenny Randles has described a similar series of seemingly baffling reports from a Lancashire Pennine region dubbed "UFO Alley" because of repeated sightings there during the 1970s, which were investigated by a Manchester-based UFO association.[22] These included reports of a swooping, silent object that frightened farmers.

Although civil aircraft movements are governed by strict regulations, military aircraft have no such restrictions and can fly as low as 200 feet in certain circumstances. Some sightings during the winter of 1987–8 were clearly the result of covert military aircraft exercises, despite the routine denials issued by civil servants at Whitehall. In January 1989, a Ministry of Defence spokesman stated in a letter that "Our low flying is done in rural areas, especially in Scotland and Wales, but obviously we need flight corridors to get to those places … We do carry out training sorties over towns, but we do not allow our pilots to fly below 200 feet … It is impossible for them to avoid towns, although we do not fly over major cities."[23]

Unfortunately, as the military are often reluctant to divulge information relating to covert night-flying exercises (indeed, it often serves their purposes to say or admit nothing), they are unwittingly feeding the growing UFO folklore and triggering mass misidentifications fuelled by media reports. Those such as the Ecclesfield sighting are eagerly swallowed by UFO believers as clear evidence of UFOs regularly intruding into British airspace. No one contradicts them, and no investigation is regarded as necessary for many believers, who follow the dictum "Why can't we just believe what witnesses say? If they say it wasn't an aircraft, it must have been an alien spaceship."

If alien spaceships really were responsible for the sightings in 1988, then they travelled millions of light years to visit one particular mining town in South Yorkshire. The evening of February 8, 1988, was one of those nights when a whole region goes UFO crazy. It was, in fact, the first of a major 'flap' which produced more than 200 UFO reports from the north and midlands of England over the following three months. Many of these came from the former mining villages and towns of South Yorkshire's Dearne Valley. I was able to investigate more than 100 of these sightings in the field, and tape record interviews with the eye-witnesses.[24]

On two particular evenings, Wednesday, February 10, and Wednesday, April 20, 1988, for example, scores of people in the former mining town of Barnsley and surrounding area reported seeing strange flying objects carrying red and green lights, and two brilliant "searchlights." On the earlier occasion, more than a dozen people independently reported seeing a huge triangular object hovering above electricity pylons between Barnsley and Rotherham. One of them said: "The object was hanging in the sky right in front of us. It was very big, black, triangular in shape and illuminated by a lot of red and green lights. We watched it for about five minutes. When we spotted it again later it had moved and was hovering above an electricity pylon. We watched it into the distance, and noticed that it followed the path of the power-lines towards Wentworth."[25]

On the night of April 20, a UFO was again first sighted in the same area of electricity pylons, reservoirs and woodland. One of the most interesting accounts was that of a housewife, Linda Sage, who saw a strange flying object whilst returning home to the village of Broomhill along an isolated country road at 9.45 pm. She told us that her daughter first drew her attention to the UFO as it approached them from the direction of Barnsley.

"As we got nearer to it, we could see it much plainer," said Mrs Sage. "It was not a plane, as there was no noise coming from it at all. It had red, green and white lights flashing. As it got right to us, it hovered round us and came lower down. It then turned all its lights out and then put two big white spotlights on my daughter and myself. My daughter became frightened, and I said to her, 'Just keep walking as we have to get home.' When we started to walk again, these big white lights came on again. It then went towards Wombwell, and we ran towards home very frightened."[26]

Mrs Sage described the classic time disorientation (known in UFO literature as "the Oz Factor") when she mentioned that "everything seemed to go quiet when it came over. There wasn't a car passed or anything while it was coming low down and looking at us."

Soon after this sighting, a group of men leaving their nightshift at a glass factory in Monk Bretton, a short distance away, apparently saw the same object hovering slightly with its coloured lights and familiar bright searchlight. Harry Reynolds, one of the workmen, told us: "I remember saying, 'If he's going to Manchester [airport], he'll never make it. He's too low' … Then for some reason I can't explain, we stopped walking and stared in astonishment, because whatever we saw it certainly was not moving. It didn't make a sound, but then it suddenly started to move. It moved in a straight line with a series of thrusting movements, almost like a dragonfly, and disappeared in a fast, darting movement, still with no sound!"[27]

Could aircraft be responsible for these puzzling observations by genuinely independent observers, none of whom knew of the existence of others who had seen the same thing? It certainly seems likely that anything which flies around so blatantly displaying red, green and white lights must be a conventional aircraft of some description. The light configurations described by these witnesses are similar to the standard ICAO (International Civil Aircraft Organisation) lights used for night flying, which include rotating red beacons and flashing high-intensity white strobes which can give a very strong flash and beam effect when seen under unusual circumstances. However, our investigations failed to trace any such aircraft or helicopters. We were subsequently informed by the AS2 in Whitehall that "We are not aware of any flying operations in the area … although we cannot discount the possibility that there may have been the odd individual training sortie here and there."[28]

But what aircraft, civilian or military, is capable of hovering motionless for extended periods of time and then move off, darting around the sky like a dragonfly without making any audible engine noise? What aircraft is capable of swooping low and projecting bright spotlights onto startled observers on the ground?

In May 1988, another 'flap' of sighting reports took place in the Midlands, where the *Staffordshire Newsletter* featured the stories of local people who had seen strange lighted triangular objects in the night sky. These reports generated so much interest that Bill Cash, the MP for Stafford, put out a request for witnesses to send details of their sightings directly to him, so that he could "put the strange sightings to the Ministry of Defence for an official investigation." The Staffordshire witnesses emphasised that these objects were "completely silent" and performed manoeuvres impossible for conventional aircraft. For example, a group of five friends in the Fernwood area of Stafford gave a description similar in some respects to the reports from April 20 in the Barnsley area of South Yorkshire.

"We were standing in the garden at 10 pm," said Eileen Ballard, "when two spotlights came through the sky towards us so that we couldn't see anything else ... They banked over and went into side-by-side formation, one above the other, and flew slowly across the sky without any noise at all ... They flew very close together and were triangular in shape, too high for microlite aircraft and too low for normal aircraft. There were a lot of lights underneath, red and green around the edge, and within that lots of others."[29]

Once again, however, a prosaic explanation for this seemingly inexplicable sighting was forthcoming after diligent and persistent inquiries by objective investigators. They found the 'UFOs' were conventional aircraft, in this case "two VC-10s of 101 Squadron from RAF Brize Norton" which were taking part in what was officially described as "an in-flight refuelling exercise" over the West Midlands at precisely 10 pm on May 16. One witnesses who was able to confirm this explanation was John Teasdale, an aircraft enthusiast, who saw the 'UFO' himself whilst driving home with his wife from Stafford to Burton-upon-Trent that night.

"We were completely baffled by what we saw until the lights (quite complex but in two clusters) were immediately overhead," said Mr Teasdale. "... then I suddenly realised I was watching two VC-10s, one immediately behind the other, the lights of the second aircraft illuminating the four tail-mounted jets of the first, and the sound of jet engines could be heard for a brief moment."[30] Although an experienced observer of aircraft, Mr Teasdale said he had never before seen such an unusual display of lights in the night sky, adding that "... it is not difficult to understand why other people had written to the newspapers to report UFOs. I can appreciate the mystery that ensued."

The refuelling exercise actually took place over the North Sea to prepare the allies for long-range bombing missions to the Middle East. No refuelling is allowed over land due to the risk of collision but witnesses saw the aircraft formation as it lined up. It is possibly more than relevant that in 1989 very similar sightings occurred over Belgium involving objects again heading towards the North Sea (see p.186).

This underlines the fact that it is very difficult for people with little or no experience of observing aircraft at night to judge height and distance of lights in the sky. In the Staffordshire and South Yorkshire examples described here, aircraft flying very high and slowly were misinterpreted by observers on the ground as low-level, silent objects which even appeared to hover. In the face of such misperception, how can any UFO report be taken at face value, particularly if it has not been subject to an immediate series of checks with local airports and weather stations?

Remotely piloted drones

In other cases, detective work has led me to suspect that some kind of low-flying, short-ranged reconnaissance aircraft (known to experts as an RPV, Remotely Piloted Vehicle, or UAV, Unmanned Air Vehicle) could be responsible for certain of the more puzzling sightings. Indeed, several of the witnesses who reported UFOs over Yorkshire that winter told me they suspected they had seen secret military aircraft undergoing tests. Two Rotherham men who saw a huge "cross-shaped" UFO hovering above their car noticed what they said resembled tiles, similar to those used on the Space Shuttle, attached to the underside of the craft. One of them, puzzled by the total lack of sound, asked me, "Are the Ministry of Defence using an unusual type of aircraft that we don't know about?"[31] Another puzzled witness from Barnsley said the object which passed over his head at low-level "was about the size of a small bus, but it made no sound: I read a lot of sci-fi and a drone sprang to mind immediately."[32]

Remote-controlled drones such as these are now used widely by the British Army in great secrecy. A United States Army report of 1983 predicted that in the near future "remote-controlled flying saucers and robot-guided vehicles will be used on battlefields." This prediction came true in 1990s when UAVs were widely utilised by the US forces in Iraq and Serbia. RPVs are also designed by commercial businesses such as Dragon Models, based in Wrexham, North Wales, who produce 25-feet-long model aircraft fitted with cameras which are used by local councils, the police and the Ministry of Defence.[33]

One UFO report which reached us during the 1988 wave of activity appears to confirm suspicions that RPVs may lie behind a number of 'unexplained' UFO reports. At 8.30 pm on April 20, 1988, over one hour before the outbreak of UFO sightings in the Barnsley area of South Yorkshire, two witnesses walking across an area of farmland at Staincross to the north-east of the town observed what they later described as "a large, triangular-shaped flying object" with coloured lights, which seemed to be operated by a man in a car who used an aerial or control box to guide its flight. At the same time, a convoy of four or five cars appeared on the lane nearby and "looked as if they were chasing this object," for the vehicles appeared to be shining their lights into the fields where the object was hovering. These cars disappeared at speed in the distance, apparently in pursuit of this "flying triangle," the man with the aerial having departed at the same time.[34]

It appears most likely that the UFO observed at Staincross – and perhaps later in the evening on the outskirts of Barnsley – was a remotely-controlled flying object. Who were the operators? What was the purpose of the exercise? And why did it apparently take place in great secrecy? One possibility is the claim made by a Mansfield-based aviator. He said that local model aircraft enthusiasts built a remote-controlled airship and deliberately made it resemble a UFO. They enjoyed the fun of inspiring yet another UFO flap!

The silent Vulcan

The description of the flying object as triangular provides another clue towards the origin of these flying objects as for decades a strange diamond- or triangular-shaped UFO has been sighted in Britain, Europe and the United States. Waves of sightings of triangular-shaped UFOs are a relatively new development both in the US and Britain, where the cultural archetype of the 'flying saucer' has become ingrained upon the collective psyche.

In more recent years, UFO believers have tried to suggest the triangular craft could not possibly be of earthly manufacture because the technology they display appears to be infinitely superior to that of our military. As a result, 'flying triangles' have become another item in the extraterrestrial belief system, which has associated these craft with such disparate phenomena as 'alien abductions' and even animal mutilations.[36] But researcher Tim Matthews has convincingly argued that a majority of the sightings of mystery triangles are caused by the testing of secret prototype Stealth-type aircraft produced by black projects, such as the mysterious Halo (high altitude/low observability) which many aviation experts suspect has been developed by the MoD at British Aerospace bases in North-west England.[37]

In many other cases, witnesses who claimed to see huge triangular-shaped UFOs have simply misidentified conventional aircraft. For example, in February 1998, a family driving home in Rotherham reported being pursued by a "triangular-shaped" UFO covered in multi-coloured lights, which "darted around the sky."[38] Checks with Air Traffic Control at the newly-opened Sheffield Airport found the sighting coincided with the approach of a **KLM** flight from Amsterdam, which had commenced its approach to the runway at precisely the same time the UFO was spotted. Subsequently, several other motorists reported seeing giant triangle UFOs hovering above the M1 near the runway, unaware that a new airport had recently opened at Tinsley nearby.

There have also been frequent reports from the East Midlands since the 1970s of what has often been called the "Silent Vulcan" owing to its shape, enormous bulk, metallic-looking construction and amazing silent flight. A large

---

THE KECKSBURG UFO CRASH

Just before 5 pm on December 9, 1965, 'something' flashed through the skies over Canada and North America. It was seen by thousands of people. Shortly afterwards, a resident of Mount Pleasant, Pennsylvania, reported seeing a "star on fire" drop into woods at nearby Kecksburg.

Locals rushed to the woods, several seeing an object about four metres in length, acorn or bell-shaped, and bearing writing like ancient Egyptian hieroglyphics. A few claimed to have seen the object glowing or giving off sparks and flames. The woods were immediately cordoned off by state police with military personnel in attendance, and the hollow containing the object isolated by armed guards.

Some people recalled seeing men arriving dressed in radiation suits before the object was removed on a flat-bed truck amidst high security. On December 12, a building contractor delivering "special bricks" reported seeing a bell-shaped object in a hangar at Wright Patterson Air Force Base.

Is it real?
● Many people claim to have seen an object in the woods at Kecksburg.
● There was a well-attested military operation to retrieve something from the woods.
● Freedom of Information documents mention "an object that started a fire."

Is it solved?
● Russian satellite Cosmos 96 re-entered Earth's atmosphere on December 9, but at 3.18 pm.
● Most of the witnesses who claimed to have seen an object retrieved did not come forward until almost 25 years after the event.
● A bright bolide meteor was also seen over North America at 4.44 pm.

Conclusions
Bolide meteor, Cosmos 96 or genuine UFO retrieval, the Kecksburg case is informative because it shows how the crash of any unknown object undergoes the same processes in UFOlogy.

Unusual markings on the object become 'hieroglyphics,' alien text if a UFO, Russian writing if the object was a satellite, or pure imaginings if it was a meteor. The usual array of crash retrieval elements, such as flat-bed trucks, tarpaulin-covered objects in hangars and dire threats to civilians, enter the narrative yet no hard evidence is forthcoming. In this case, all the evidence points to the mysterious object being a bolide meteor.

Further reading
A History of UFO Crashes. pp. 95–120. Randle, K. Avon, 1995.
UFO Retrievals. pp. 102–108. Randles, J. Blandford, 1995.
"Incident At Kecksburg." pp. 9–10. *Unopened Files.* Quest Publications, 1999.

---

percentage of the UFO sightings reported during the wave of 1987–8 describe large, triangular-shaped flying objects carrying rows of coloured lights. The lack of sound is probably the most distinctive feature of these sightings. Whenever detected, it is usually only when the witness is within close proximity to the flying object. The noise is often described as a soft humming or buzzing, like a swarm of bees or an electric motor. A classic sighting describing this type of UFO was made on September 22, 1987, near Abbots Bromley in Staffordshire. One of the witnesses, Dominic Goodwin, who was 22 at the time, gave the following vivid account of his experience:

"My family and I were driving back from Walsall illuminations and had just passed through Baggots Wood at Abbots Bromley when my father shouted, 'What on earth is that?' When I looked up, I saw the underneath of what seemed to be a triangular-shaped craft. The craft was moving very slowly, almost stationary. I wound down my window and there was no noise at all. The craft slowly banked over to one side and I caught a glimpse of the top of the craft. It was slightly raised on one side. The underneath of the craft had lights which were pulsating in sequence. Down the middle of the undercarriage was a red cross-shape, very dimly lit. On each point of the craft were like large strobe lights. The craft moved away slowly towards Abbots Bromley. It then stopped over the woods, and all I could see was a long, thin red stripe in the sky directly over the woods. I would like to state that never in my life have I ever seen such colours as I did on the craft. It was truly an unforgettable experience."[39]

In the late 1980s, newspapers and aeronautical magazines had been speculating for some years about the the existence of the previously 'top secret' Stealth fighter. Rumours about the Lockheed F-119A had spread because although the super-secret craft is supposed to have been operational since the late 1970s, no photographs of the craft were released by the Pentagon until 1988. It is clear that ground observers who see an aircraft of this kind flying at night would perceive it as triangular-shaped. The engines are positioned deep inside the airframe to reduce noise and heat emission, explaining the curious lack of noise reported in some instances.

On November 10, 1988, the US Air Force for the first time confirmed the existence of the F-117A. According to Pentagon sources, pictures of the previously classified aircraft were being released because the USAF "... wants to operate the type from a wider range of bases and also during the day; previous flights have been made almost exclusively at night."[40] The Pentagon revealed that the Stealth fighter first flew in June 1981 and has been operational since

October 1983. It later became clear that a number of the original 52 aircraft had visited, or were currently based, in Britain as part of the special partnership between the two countries under the umbrella of NATO.

In fact, there had been persistent rumours that Stealth and other experimental aircraft were operating out of a number of UK airbases under great secrecy.[41] In September 1994, there were reports that one of these prototype triangular aircraft, possibly a TR-3A Black Manta, crashed on the runway at RAF Boscombe Down in Wiltshire. The twin-tailed aircraft was quickly removed by the military, but not before it was spotted by aircraft enthusiasts.[42] There had been several sightings of hovering UFOs in the area of the base. These have already been used to suggest the object which crashed was of extra-terrestrial origin. This kind of rumour works to ensure that prototype aircraft secrets are carefully hidden behind a smoke-screen of disinformation.

Pilot magazine reported in April 1986 that the F-117A had been developed from an earlier Lockheed design known as XST (Experimental Stealth Technology), and that it was capable of flying up to speeds of 1,500 mph. The shape of the aircraft had been deliberately chosen to make it almost invisible to enemy radar, aided by a 'chameleon' skin-coating known as RAM (Radar Absorbing Material), a material manufactured in the UK which, under computer control, enables its surface colour to adapt to the surrounding terrain. The F-117A is known to have been responsible for a Californian UFO report in 1975, when a radar unit attached to Edwards Air Force Base tracked a target travelling at 460 miles per hour which vanished in one sweep from the radar screen. The Stealth had, in fact, switched to its 'invisible mode' shortly after it appeared on screen. William Spaulding notes that "Officially the report was logged as 'unidentified,' one part of the USAF seeing no need for another to know what it was doing. The foundations of the UFO myth were made a little stronger – and the secret aircraft remained secret."[43]

This large body of convincing evidence leaves little doubt that aircraft – including secret prototype craft – can explain many UFO reports, particularly those concerning triangular and delta-winged objects. How many classic UFO sightings of the last two decades may be misidentifications of 'secret' military technology, prototype Stealth aircraft, Remotely Piloted Vehicles and other unusual aerofoil configurations such as giant airships? The advantage provided by the widespread will to believe in alien UFOs on the part of the public may therefore provide the perfect cover required for the clandestine testing of military hardware in unrestricted areas. It is known, for instance, that the Russian authorities have used UFOs in the past as a cover-story for top-secret launches of rockets, and it is highly likely that intelligence services in the west have been equally interested in the use of UFOs for psychological warfare purposes.[44]

The UFO wave of 1987–8 is good evidence of how the modern myth of alien visitors is created by the popular media, and transmitted into reality through wishful thinking onto that giant Rorschach blot called 'the night sky.' The misidentifications of celestial phenomena and conventional aircraft revealed by these investigations demonstrate how the public is largely ignorant of astronomy and perceptual psychology. Combined with an awareness propagated via media hype that others have recently reported UFOs, and fuelled by sensational tabloid headlines, many people are led to make a false interpretation of a light or object seen in the sky they might otherwise have ignored. Why else would otherwise sensible observers interpret a bright white light which remains stationary in the sky for more than four hours as anything other than a bright star or planet?

This conclusion appears logical to us, but psychologists would argue that UFO waves are produced as a result of 'collective wish fulfilment.' As sociologist Dr Phyllis Fox concludes "... the process by which people arrive at beliefs about the nature of UFOs is similar to the process of rumour transmission."[45]

Many UFO waves and individual 'encounters' are created in the same fashion today, but in the late twentieth century they have taken on all the symbolism of a culture obsessed with aliens and the conquest of space. During the course of a century, secret inventors and benevolent space brothers have been replaced by sinister 'greys,' but all of these alien types are merely reflections of our wishes, fears and hopes as mankind prepares to venture into the vast emptiness of space. To paraphrase the sociologist and psychologist quoted at the opening of this chapter, the answer to the UFO mystery is not likely to be found by scanning the skies with telescopes. It will ultimately be discovered by understanding the inner workings of the human mind to answer the question of why the symbol of the 'flying saucer' has captured the imagination of the world.

5

BETWEEN A ROCK
AND A HARD PLACE
The Cracoe UFO photograph

ANDY ROBERTS

"A BUFORA investigator has stated that the Cracoe photograph is 'light reflection.' What UTTER RUBBISH!" – Mark Birdsall *YUFOS Journal*, November 1983.

"The Aliens Have Landed" screamed the *Daily Mirror* for Tuesday, August 23, 1983. The headline was unequivocal, the accompanying photograph strange. It showed a bright, apparently luminescent, phenomenon against a rock face in the Yorkshire Dales. Journalist John Gapper was impressed, writing, "This is the picture that UFO spotters claim is 'definite proof' that alien beings have landed on earth," before going on to reveal how two policemen had watched the UFO "hovering above the ground for an hour."

Graham Birdsall, president of the Yorkshire UFO Society (YUFOS), was quoted as saying: "This is definite proof of the existence of UFOs. I am very excited about the photograph. It has convinced me that this is an area which they visit regularly. Something mysterious is happening out there, but I don't know what it is."[1]

This newspaper article was the first major revelation to the British public of what came to be known as the "Cracoe UFO Photographic Case." The photograph was already two years old in 1983, but the case was to rumble on for another four years before it was finally laid to rest. And repercussions from the investigation are reverberating through the British UFO research community to this day.

Most photographs of alleged UFOs can usually be dismissed after analysis as obvious hoaxes, or camera or film-processing faults. UFO buffs place great store in any photographs which elude explanation. Belief is strong that if a UFO is captured on film, it is somehow representative of the subject's reality status, elevating it from the shady realms of misperception or fantasy into those of scientific proof. However, less than critical UFO investigators are often quick to seize an unexplained UFO photograph in order to back up their particular beliefs and theories, "The camera cannot lie," being their argument.

The reality is usually somewhat different. A far more apt quote when dealing with UFO photographs is not "The camera cannot lie," but "Every picture tells a story." That the perception of unexplained photographic cases frequently leads to confusion and controversy is exemplified by the story which unfolded in the wake of the Cracoe Fell photographs.

The hamlet of Cracoe is situated just to the north of Skipton in the Yorkshire Dales National Park. It is a small place, with only a few hundred inhabitants, boasting a scattering of farms, houses, a pub and at one time a police house where the local constable lived. This police house and its occupants were the focus for the initial events of the Cracoe sighting on March 16, 1981.

At 10.55 that morning, Police Constable Steve Guest's wife was in the kitchen preparing a cup of tea for her husband and a visitor, PC Derek Ingram. The kitchen window overlooked a wide sweep of farm and moorland, which eventually rises steeply to the summit of Cracoe Fell. As she gazed out across a landscape she and her husband must have seen many times each day, Mrs Guest's attention was drawn to an amazing sight. According to the YUFOS file on the matter, "her eyes locked on to a most incredible sight. Suspended around the central rocks at the top of Cracoe Fell a series of five brightly lit spheres were shining. The glowing orbs hurt the witness's eyes. For two minutes, she remained at the window trying desperately to solve this most puzzling sight. She could not." Mrs Guest immediately called her husband into the kitchen to witness the puzzle. His official statement bears close scrutiny:

"On Monday 16th March, 1981, I was in the kitchen of my house when I saw three bright lights on the rock face. I looked at the lights with my binoculars, but was unable to get a clear view, as if I was looking straight at bright car headlights. The lights were in a line, but there appeared to be a smaller light just to the side of the main light source. I saw a shape at the back of the lights, but was unable to make it out.

"Sometime around 11.30, two RAF jets flew over the area. First one, then another, they were flying very slowly. There is no water on the rock face to give such a reflection, and to the best of my knowledge no metallic deposits. At about 11.55, the lights dimmed and became bright on several occasions, before disappearing. I have observed the fell every day at the same time. The light source has not reappeared. I have also spoken to several village residents who state they have no knowledge as to what could have caused the lights."[2]

PC Ingram was called into the kitchen to witness the amazing sight. His account amplifies PC Guest's:

"Please see account of officer Guest, which I totally agree with. I remained in the house during the entire incident until the lights disappeared. I am a keen amateur photographer and took the pictures between 11.15 am and 11.40 am. The lights appeared just below the top of the fell: they were hovering. The array of lights varied in intensity, and after one hour they vanished. There was no concrete shape yet the colour was the same as magnesium lights; they were brilliant. I found the lights unusual because on the fell (which is very steep) there is nothing to stand the lights on. The terrain makes it impossible to duplicate such an event."[3]

That these witnesses to the Cracoe Fell 'UFO' were insincere in any way was never an issue. Nor was there any dispute about the veracity of the photographs which were taken. They were emphatically *not* hoaxed. But neither witness testimony nor photographic imagery is proof of any objective reality. However, it was clear that these two police officers, trained in observational techniques and familiar with the local terrain, were spellbound by what they saw. There can be no doubt from both their statements and photographs that they believed *something* strange was taking place in Cracoe that day.

The phenomenon was clearly dramatic and memorable. So much so that when Derek Ingram recalled the event for a 1996 Discovery Channel TV documentary it was still vivid in his mind – "I saw a very intense bright light, a big band of light on the rock face of the fell."[4]

At 11.15 am on March 16, partway through the sighting, the officers telephoned their sergeant, Tony Dodd, who was based in nearby Skipton. Dodd was a keen UFO buff and editor of YUFOS' Journal, *Quest*. He assured the officers that he was on his way, but that they should take no action as yet.[5] At any time during the sighting, it would have been a relatively simple matter for the officers to drive halfway up the fell and walk the rest of the way to determine the cause of the phenomenon. People often query why they did not. But the witnesses had no idea how long the sighting would last, or how important it was going to become for believers in Unidentified Flying Objects.

At 11.20 am, one of the officers made a "vital discovery," seeing a "triangular fin" behind the central light. A 'fin' seemed to imply attachment to an object. What – or who – on earth could be responsible for a brightly lit object hovering against the side of a sheer rock face?[6] This was the first hint that the Cracoe UFO may have been an artificial construction.

Right: This 'crop' circle, found in snake-infested reed beds near Tully, Queensland, Australia, in January 1966, inspired British hoaxers to imitate.

Right: How the media exploits mystery and enforces images of aliens. This fake circle – not created by the TV company – is being filmed for a news programme with an actor dressed as an alien.

MEXICAN WAVE (p. 196)
Right: The first known UFO photograph taken by an astronomer through his tele-scope in Zacatecas, Mexico, in 1883. It is believed the 'UFO' is an insect on the lens!

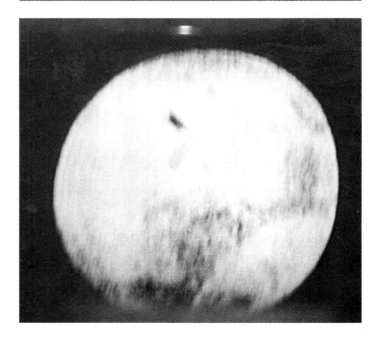

PERFECT CASE? (p. 52)
Above: Witness Peter Day

THE WILLAMETTE PASS PHOTO (p. 103)
Above right and right: The 1966 photo at
Willamette Pass, Oregon. The disc-like UFO climbs
from the trees but is later found to have been a
road sign being passed at speed. (Irwin Weider)

Below: A typical photographic accident which can
fuel UFOlogical rumours and claims. This image,
created in Salem, Massachusetts, shows lights
reflecting in the window through which the photograph was taken.

At 11.55 am, the UFO was still being observed by the police officers. Then its lights began to pulsate in unison, dimming and brightening several times until, one by one, they went out and the UFO had vanished. Shortly afterwards, the astonished officers watched as two military jets "crossed right over the spot where the unknown targets had hovered ... The officers were puzzled by this."[7] The bizarre sighting had lasted almost an hour. During that time six colour transparencies were taken of the phenomena. Unfortunately, both police officers were off-duty. Had they not been, an official police enquiry into the matter may have solved the case before it assumed mythic proportions. But clues as to the true nature of the case were seeded very early on in the investigation. On the first day, in fact.

Local farmer Derek Carlisle had been outside the police house during some part of the sighting. He told the officers the phenomenon was not a UFO, but a bright rock reflection, one he had seen many times before. The police witnesses dismissed his claims out of hand. YUFOS claimed they had evidence which "negated" Carlisle as a "vital observer," but agreed that as he had been looking at the same phenomena as the police, his theory should be taken into account.

The police officers made statements shortly after the UFO had disappeared and the slide film was passed on to a YUFOS investigator for processing at a "confidential address in Hull."

In the weeks following the sighting, YUFOS investigators did an excellent job in determining what the UFO could *not* have been. The area was examined in detail. Nothing could be found to account for the phenomenon. Environmental factors such as snow, ice and running water as a possible cause for the sighting were also ruled out. YUFOS enquiries revealed there had been no helicopters in the area, but they considered the overflight of military jets at the conclusion of sighting to be "most interesting," and "more than coincidence." As there had been a major NATO exercise on March 16, it was at first thought that the lights may have been some form of target or marker. The RAF verified they had fighter jets on low-flying practice in that area, but the "object," as they put it, had nothing to do with them. In any case, PCs Ingram and Guest had gone to the top of the fell as soon as the sighting was over and found no evidence of any military hardware. YUFOS therefore concluded that "The military aircraft overflights had, in fact, seen the light formation and made a pass."[8]

YUFOS were rightly puzzled by both the sighting and photographs. But rather than release the information to the media immediately, they sensibly chose to spend two years investigating the case in some considerable depth before releasing their findings.

But for the inexplicable photographs, this event would probably have gone down as just another lights-in-the-sky-case, intriguing at the time, but impossible to verify or investigate. But photographic evidence is rare and valuable. YUFOS had the pictures analysed at four establishments, the now defunct Ground Saucer Watch (GSW) in the USA, Klaus Webner, a UFO investigator in Germany, a "police source," and at Leeds University.

Full results of the lengthy analysis from the first two were released in 1985 by YUFOS. Like the farmer's claims, this analysis flagged up points which should have been taken more seriously at the time. Ground Saucer Watch concluded that whilst they were convinced the photographs were genuine, "there is no evidence that the anomalous images are 'objects' hovering between the witnesses and the distant mountains hills."[9] Klaus Webner's conclusions included the statement, "I have found no evidence that there was anything in the air between Cracoe Fell and the eye-witnesses."[10] Webner also believed the photographs were genuine, not hoaxed, and mooted the possibility they were lights used by mountaineers. Neither report could positively identify the phenomenon. Nor did they mention a 'fin' or craft of any kind.

Far more interesting were the comments made by the analyst at Leeds University. Due to his position as the head of a science department, he requested anonymity in YUFOS' published reports. After studying the photographs in depth, he called the YUFOS investigative team together and astonished them with his conclusions. He was convinced the photographs showed something above the lights and said: "What you probably have is a most unusual structure which looks oval to the eye, yet is perhaps round. Beneath the structure (or craft) appear to be three almost circular globes ... You have an object which is tilted back towards the fell face ... It is displaying three remarkable and very luminous lights. I do not think it is natural." The pseudonymous 'Mulligan' drew a picture to illustrate what he meant. The drawing was of a classic Adamski-type flying saucer.[11]

YUFOS were "both delighted and bemused." "What," they mused, "was the scientist trying to tell us?" Whereas the GSW and Webner's analyses were open to interpretation, Mulligan's was not. Here YUFOS had a qualified scientist, working at one of the country's finest universities, effectively claiming that a flying saucer had been captured on film.

The three Cracoe Reports (published between 1985 and 1986) contained a vast amount of investigative data concerning the initial sighting, photographic analysis and theories pertaining to the case, both for – and against – the phenomenon being a structured object. But by far the most significant evidence in favour that it was came when Mrs and Mrs John Ackroyd related their sighting from March 16, 1981.

From South Yorkshire, the Ackroyds were on an excursion to the York-shire Dales, driving past Cracoe at around 3.45 pm when they saw three glowing spheres high over the fell. They initially believed them to be aircraft lights, but the apparent speed and height were not consistent with any aircraft with which they were familiar. After stopping their car and observing them for a while, it became clear there was a "dark shape" above the lights. Now they were certain it was not an aircraft. Puzzled, the couple drove off, and did not mention the incident until they attended a YUFOS lecture in 1985 in which the Cracoe case was featured heavily. At the close of the meeting, Mr Ackroyd approached YUFOS officials, saying simply, "My wife and I saw that thing over Cracoe."[12]

The implication was potentially earth-shattering. Given that the phenom-enon the police officers and the Ackroyds saw was the same, a "structured craft of unknown origin" was indeed in the skies above Cracoe. What's more, it must have been in the area, if not constantly in the air, for over four hours on that fateful day in 1981.

YUFOS came to no firm conclusions as to the specific nature or origin of the 'UFO,' but had already publicly stated in 1983 that "an unknown structure lies behind the lights – this is covered by a stream of white lights," and "for the record, the Cracoe UFO is undoubtedly solid."[13] I recall querying the nature of the phenomenon at a 1986 YUFOS lecture in Burnley and being told it depicted a "structured craft of unknown origin." It was clear that they believed it was solid, and capable of flight and light emission.

Throughout the Cracoe reports, YUFOS had often referred to and carefully considered Farmer Carlisle's contention that the UFO had its origins in reflected sunlight. By the end of the third edition of Cracoe: The Evidence, their research teams had narrowed this possibility down to zero, saying: "YUFOS research conjectured that the three almost circular spheres of dazzling light were not caused by sunlight reflection. The photographic content confirms this, as does scientific evidence."[14]

Until YUFOS released the photographs to the media in 1983, the Cracoe case was virtually unknown both to the public and within the UK's tightly-knit UFO research community. No UFOlogists from other research groups had been involved in the initial investigations, and YUFOS chose to share their evidence with only a few trusted colleagues. Once the case became public knowledge, other UFOlogists began to scrutinise the case.

The popular perception of UFO investigation is that UFOlogists share one goal, that of solving the UFO mystery, and will happily assist each other where necessary. This is a completely erroneous view of the subject. In reality, the UFO community is comprised of many small, warring factions, each driven more by belief than fact. UFO groups rarely share information, especially if one group is likely to take a different view from another.

On the surface, this may appear counter-productive, but it is actually a positive state of affairs because it is only from these frequently intense disputes and re-evaluations of evidence that the truth of a case often emerges. Had no other investigators become involved in the Cracoe case, it would probably still be listed as 'unknown' and the photographs now elevated to classic status. YUFOS investigation of the Cracoe photograph is only half the story. The denouement comes in the aftermath of the case being made public and its subsequent re-investigation.

YUFOS released the photographs to the media in August 1983. Newspapers and TV immediately seized on the images, which were featured widely in the national Press. The *Daily Mirror*, whose headline opened this chapter, faithfully promised YUFOS they would not hype the alien angle, but went ahead and did it anyway. To their credit, YUFOS had never overtly claimed the Cracoe UFO was alien in origin and demanded an apology. But it was too late: the damage had been done. Whatever anyone thought of the Cracoe UFO, in the eyes of the public it was extraterrestrial in origin.

Local newspapers played their part in spreading the Cracoe myth, too. The *Yorkshire Post*'s front page offered "UFO Over Yorks" whilst the Skipton-based *Craven Herald*, Cracoe's local paper, ran a sensible piece entitled "UFO Siting (sic) Confirmed." They quoted YUFOS Graham Birdsall as saying: "We have kept this under wraps for two years because we wanted to be absolutely certain the photograph could not be knocked down when we released it. During that time, we have covered the area with a small toothcomb, and it has left us in no doubt that no natural phenomenon could have caused this to hover there for just under an hour."[15]

It only took six days before the *Herald* was running a piece which effectively 'knocked' both the photograph and Graham's certainties for a six. The article was headed "UFO Rubbish!" and read:

"Reports that a shiny object seen on Cracoe Fell were conclusive proof of alien visitors to the earth have been dismissed as "rubbish" by a local farmer.

Hetton farmer Mr D. Carlisle said the phenomenon often occurred on dull days when the sun caught rocks on the fell. 'It's quite spectacular, but that's all there is to it,' he explained. He was present on the morning two years ago when two policemen photographed the shining fell, and recognised it as the same optical illusion he had seen there before."[16]

This account of the Cracoe UFO's origins did not fit in with the media's fondness for stories about aliens and was soon forgotten. But experienced UFO researcher Nigel Mortimer saw it. Had it not been for his vigilance, the outcome of the Cracoe case may have been entirely different.

Nigel lived quite near Cracoe and decided to investigate the case on behalf of BUFORA. After seeing the media furore about the Cracoe case and specifically Farmer Carlisle's statement, Nigel concluded that there could well be a prosaic solution to the case and began to visit the area regularly, confident the rock reflection hypothesis was valid. Although he saw many rock reflections, none was anything like the Cracoe 'UFO.' Nor could he identify the exact location of the phenomena captured on film by the police officers. But serious doubts had now been raised about the case – and they spread like wildfire through the British UFO community.

Meanwhile *Quest*, the YUFOS journal, regularly featured updates concerning the on-going Cracoe investigation. One such piece about the photographic analysis was immediately followed by an article titled: "Just Coincidence? The Cracoe Connection." This dealt with several sightings of flying discs with three bright balls on the underside, all from the Cracoe area. Was this what the scientist at Leeds University was trying to tell YUFOS? Writer Mark Birdsall concluded, "We have already had a tantalising glimpse of something very similar in relation to the argument for one particular type of unknown vehicle ... The Cracoe UFO."[17]

In March 1986, I was asked by Paul Devereux, author and then editor of *The Ley Hunter* magazine, to write a piece about the Cracoe UFO as a possible example of 'earthlight' phenomena. I had been interested in the Cracoe area for several years as a focus for earthlight activity and agreed carefully to analyse both the YUFOS data and Nigel Mortimer's speculations before coming to any conclusion. Problems were immediately encountered in obtaining a copy of the YUFOS report for research purposes. These were advertised in YUFOS' journal as being available to anyone, but my request was refused. The reason given was because "you would not agree with YUFOS' conclusions."[18] The case was getting interesting already!

Nigel Mortimer was digging deep into the case, too, and facing similar obstructions. In a heated telephone conversation with YUFOS' Director of Research, Mark Birdsall, Nigel was told he had no business investigating a case which "wasn't his." "How can anyone 'own' a UFO case?" Nigel wondered. The situation deteriorated further when YUFOS' Executive Committee issued a statement in which they disassociated themselves from national investigations co-ordinator Jenny Randles, Nigel Mortimer and the British UFO Research Association, saying they did not regard them as serious investigators. This followed BUFORA's open support of the light reflection theory.[19] It was clear that sceptics were not going to enjoy any co-operation with YUFOS, as they clearly regarded the Cracoe case as 'theirs.'

This was just the motivation we needed to persist in cracking the case wide open. I and other researchers from the West Yorkshire UFO Research Group (WYUFORG) began to visit the Cracoe area frequently, searching for clues to the mystery. But we were in a difficult position. Unable to obtain access either to the original photographs or to the main witnesses, we only had YUFOS reports to go on. And YUFOS would not move from their certainty that the Cracoe UFO photographs represented a "structured craft of unknown origin."

Intrigued by YUFOS' continual refusal even to consider, never mind actually discuss, an alternative explanation for the Cracoe 'UFO,' we decided to contact the farmer, Derek Carlisle, ourselves. He was only too pleased to be interviewed. In September 1986, we spoke to him for an hour at his farm in the village of Cracoe. Mr Carlisle stood by his 1981 comments to the local newspaper saying:

"I was present outside Cracoe police station on the 16th of March 1981. I observed the lights for not more than 15 minutes. The lights were on Rylestone Fell ... The weather conditions were overcast with outbreaks of sun. The lights I observed were as portrayed in the photograph and in that location. I have seen these lights both before and after on many occasions, as have my wife and son. The lights appear when the rocks are wet ... and when the sun shines on the wet surfaces ... My attitude towards the UFO phenomenon is one of an open mind. In my opinion, the lights I saw were nothing else other than the sun shining on the rocks. On the day in question, the lights were brighter than I've seen before. I did not notice any structure whatsoever behind the rocks."[20]

Mr Carlisle also noted that he, like many Cracoe villagers, thought there was – and is – "something" unusual going on in the area, adding that a few weeks before our interview, bright lights were seen at night high on Cracoe

Fell. The fact that he was not totally sceptical of the UFO phenomenon as a whole gave further credence to his story, even though YUFOS, who had spent hundreds of man hours and thousands of pounds on the case, still vehemently disagreed with him.

If Mr Carlisle's version of the sighting was correct, the phenomenon could theoretically be treated scientifically and replicated. Replication of the original photograph would be proof positive of the true nature of the Cracoe UFO. The only problem was just how to catch the phenomenon in action. I and other WYUFORG members literally 'stalked' the fell, visiting it several times a month at all hours and in all weathers throughout the autumn of 1986.

THE WILLAMETTE PASS PHOTO

Whilst driving over Oregon's Willamette Pass during November 1966, a college science professor took several photographs of the snow-covered landscape. As he took his last picture, he saw a UFO through the view-finder for a few seconds. The processed film revealed a curious three-tiered, disc-shaped object.

The UFO seemed to be rising from the ground with some kind of propulsion emitting from its underside. One investigator suggested it was sucking up snow as it rose from a forest clearing. The unusual nature of the photographed object puzzled the UFO community and various theories were put forward to account for its apparent behaviour. Perhaps the strangest theory was that the UFO had phased "in and out of our reality" three times in less than a second to cause the triple-disc effect. As late as 1992, the photograph was still accepted as a genuine UFO.

Is it real?

● The photographer was a PhD scientist who never sought to use it to make money.
● He had seen the object through the camera's view-finder.
● UFO expert Adrian Vance calculated the UFO was seven metres in diameter.

Is it solved?

● Physicist Irwin Wieder found inconsistencies between what the photographer said and what appeared on the film.
● After visiting the location and seeing a roadside marker, he believed the image could have been a snow-capped road sign.
● Wieder duplicated exactly the sign from Willamette Pass and photographed it on site from a moving car – and the photographs were almost identical to those taken in 1966.

Conclusions

The Willamette Pass photo is a classic example of how a mundane object can be photographed and misperceived as a UFO. Instead of performing in-depth analyses, UFOlogists accepted the photograph as genuine and sought bizarre theories to explain their misperceptions. The professor's scientific credentials were frequently invoked to 'prove' that he could not have been mistaken or a hoaxer. But it took good fortune, persistence and a remarkable theory to finally solve this case.

Further reading

UFOs And How To See Them. pp. 61-63. Randles, J. Anaya, 1992.
"The Willamette Pass Photo Explained." pp. 13–15. Wieder, I. *UFO Magazine,* Vol. 12, No. 6. Jan/Feb 1994.

During this period, YUFOS published an issue of *Quest* which reviewed the evidence to date and railed against those investigators who had dared to voice an alternative opinion about the case. It highlighted a letter from one of the original witnesses, PC Derek Ingram, in which he reiterated his beliefs about the sighting, claiming, "In my opinion there was an unexplained object on the fell on the day in question, and how anybody can put it down to mere reflection is beyond me."[21]

November 1986 finally brought a breakthrough. On a family day out in the Yorkshire Dales, I stopped in Cracoe as usual to look up at the fell, more out of habit than hope. The weather was cloudy with occasional shafts of sunlight. I stared at the distant rocky ridge, yet again pondering just what could have been so dramatic enough to enchant two police officers for almost an hour. As I mused, my attention was caught by a glint of light. Attention slowly turned to interest and then rapidly to astonishment as I realised I was looking at the Cracoe UFO! Just below the ridge of the fell, some two miles away, was a narrow strip of brilliant light, interspersed with bright 'blobs.' I grabbed the binoculars from the car and studied the light closely. There was no doubt about it: this was the Cracoe 'UFO.' Had the "structured object of unknown origin" returned five-and-a-half years on especially for my benefit, or was I looking at the elusive rock reflection seen by Derek Carlisle and speculated on by Nigel Mortimer? Was I looking at the *real* Cracoe UFO?

I returned to Cracoe a week later. Now I knew exactly where to look, the 'UFO' was instantly visible, albeit again not as bright as the 1981 police photographs. Several shots were taken with and without a zoom lens. When these pictures were processed and the image enlarged, the true nature of the Cracoe 'UFO' was apparent.

Comparison done by Mike Wootten for BUFORA showed that the two images – the police officers' from 1981 and mine from 1986 – were of *exactly* the same phenomena. Now I had identified the correct area on the fell, I visited the site to see the now-landed Cracoe 'UFO' at first hand. As Farmer Carlisle correctly believed, the 'UFO' was due to an optical illusion. The cause of all the fuss was a rock surface just below the ridge of the fell, easily accessible by foot and about a forty-five-minute walk from the roadside. Water draining from the moor combined with lichen to produce three vague marks on the angled rocks. This combination of circumstances, coupled with tiny quartz crystals embedded in the gritstone rock, gave off a reflection which appeared as a strip of white, with three or more circles in it. The reflection could only be seen clearly from specific locations and under certain light conditions, although once the viewer knows where to look it can be identified at any time in daylight hours.

But why had no one cracked the case before? YUFOS unwittingly obfuscated independent researchers by printing the police officer's photographs the wrong way round. This made it extremely difficult to locate the exact area of fell amongst the tumult of rocks on Cracoe Fell. In addition, the photographs used by the media and in YUFOS publications and lectures were enlargements. This made the 'UFO' appear considerably larger than it actually was. To the naked eye, the phenomenon, no matter how bright, is quite small and distant when seen from Cracoe village. This was never made clear by YUFOS investigators, who had also insisted their calculations gave the length of the 'UFO' as over ten metres. In fact, the entire rock on which the 'Cracoe UFO' appears is barely five metres long. YUFOS investigators must have walked past, if not actually over, the 'UFO' on several occasions!

Several UFO researchers from other groups were shown the photographs and taken to the fell to see the phenomenon at first hand. All agreed that the 'Cracoe UFO', the "structured object of unknown origin," was nothing more than an unusual rock reflection.

I then contacted one of the original YUFOS analysts, Klaus Webner, and sent him copies of my photographs for his comment. He replied: "The 'Cracoe UFO' is unmasked. Your slides show the same phenomena under controlled conditions. The intensity of the light is not the same as it is on the Cracoe photos, but the position of the reflection is absolutely the same. Your slide is evidence that a harmless reflection on this sloping area of rock is responsible for the huge UFO headlines in the newspapers."[22]

Now completely certain I had caught the Cracoe 'UFO' on film, the most logical course of action was to inform YUFOS, to give them the opportunity to examine the evidence and retract their grandiose statements about the photographs. Despite their reluctance to share information, WYUFORG and I had consistently operated a policy of informing YUFOS officials of our every move in attempting to solve this case. Philip Mantle, their Overseas Liaison Officer and "team leader on ground research" in the Cracoe investigation, was appraised of the situation and invited to view the proof. He declined, saying that the next issue of *Quest* would "leave you and your colleagues in no doubt that the Cracoe photographs do not depict light reflection."[23] It did not!

YUFOS punctuated their head-in-the-sand attitude by issuing yet another edition of the Cracoe report. Hoping this might at least shed new light on their reluctance to accept alternative evidence, I attempted to obtain the report, only to be told, "As you are probably aware, liaison with WYFORG does not exist, and to allow reports which are for the benefits of serious researchers to be sent to your group at this time will conflict with our current attitude towards your group."[24]

YUFOS might not have been interested in the reality of the Cracoe 'UFO', but the *Yorkshire Post*, which had broken the story in 1983, was. Journalist Tim Zillessen was fascinated by the twists and turns of the case and quoted me as saying: "We believe we have incontrovertible proof that it is nothing more than a complex light reflection. Undoubtedly, a lot of people saw something that day, but unfortunately they do not accept a rational explanation for it and still refuse to do so. We did not set out deliberately to dispel or to disprove the sighting: we simply set out to investigate it. We are open-minded enough to accept a UFO sighting when it happens, but not in this case."[25]

In the interests of balance Zillessen contacted YUFOS for their comments. Mark Birdsall, still refusing to accept the case was solved, opined: "We absolutely reject any suggestion that the sighting was a light reflection. We are convinced something was seen on that day on the fell." Birdsall attempted to divert *Post* readers from the facts by making "a stinging attack on the research group and the photographic analyst. He said the group had only been in existence for three years and did not have enough information to make any positive claims. He dismissed the analyst as a great sceptic who had no scientific authority to make any judgements." Obviously the scientific replication of the exact phenomena did not count as "enough evidence" for YUFOS to retract their claims![26]

Mark Birdsall claimed he had again visited the officers who witnessed and photographed the 'UFO' and they still stuck to their accounts. Whether the original witnesses stuck to their accounts or not was yet another red herring and never an issue. It was the interpretation of the phenomenon caught on film which had been under scrutiny. In a final attempt to convince YUFOS, Graham and Mark Birdsall, together with Philip Mantle, were invited to see the photographs I had taken of the Cracoe 'UFO'. The meeting was not a success. It appeared that YUFOS could not – or would not – accept the factual evidence before them.

However, the majority of UFOlogists chose to believe the scientific interpretation: that the police had photographed and misperceived a rock reflection.
The only alternative explanation, if the 1981 police photographs *did* depict a
"structured craft of unknown origin," was that this particular 'UFO' enjoyed the
ability to replicate the exact size, shape and position of a naturally occurring
light reflection. What are the chances of that happening?

Although the case was now solved, huge question marks still hung over the
whole affair. Why did two police officers become transfixed for almost an hour
by a tiny natural phenomenon? Why did the case attain the status of a cause
célèbre for YUFOS, being featured heavily in their magazine and becoming a
centre-piece of their lectures? Why did a scientist at Leeds University claim the
photographs depicted a flying disc? Why did the Ackroyd family believe their
aerial sighting later that day was the same phenomenon as shown on the Cracoe
'UFO' photograph? These are just a few of the questions raised by the Cracoe
case. The answer to them all can be summed up in two words – expectation and
misperception.

The misperception in the Cracoe case was truly dramatic. The police officer
who lived in the police house at Cracoe had done so for quite some time.
According to their sergeant, Tony Dodd, "his knowledge of the location is
second to none." Both he and his wife must have looked across at the fell from
their kitchen on numerous occasions without seeing the 'UFO.' Yet the combination of environmental phenomena which comprise the Cracoe 'UFO' is
visible to some degree or other *every single day*. Why – and how – on the
morning of March 16, 1981, did the natural become the supernatural?

Unusual and unidentified lights *had* been seen in the Cracoe region for a
number of years prior to 1981, possibly for centuries if folklore is to be
believed. Some locals apparently even referred to the area as "flying saucer
alley," so it could be said there was a long-standing *tradition* of UFOs in the
area. Since the late 1970s, the Yorkshire UFO Society had been quite active in
the area, interviewing witnesses and holding skywatches. Their presence and
beliefs were reported by the media, so the UFO myth was fed back into the
community, and so on. UFO sightings were part and parcel of the contemporary, living folklore of the area in the 1970s and '80s.

This tradition was well known to the police witnesses involved through their
sergeant in Skipton, Tony Dodd. In his book *Alien Investigator*, Dodd recounts
how he saw many UFOs in the Cracoe area from 1978 onwards. Several of
these were witnessed whilst in the company of other police officers. Many were

of quite dramatic craft. One which Dodd saw only miles from Cracoe was a "large object, dome-shaped, with white light coming from what looked like windows," whilst another from the same area was "a massive disc, with a dome shape on top."[27]

Dodd's belief was that these craft were alien in origin and that his sightings were "an education I was being *given* by the aliens."[28] It is likely that Dodd's interest in UFOs, if not his belief in aliens, was widely known throughout the police force in the Cracoe area, and that this led to a cross-contamination of belief and expectation. Because of this climate of expectation, a lump of shiny rock became misperceived as a UFO.

The subject of misperception is central to any understanding of the UFO enigma. But it is widely misunderstood. Most people cannot comprehend how even the most highly trained observers can misinterpret mundane objects as UFOs. But they do. Each year thousands of people see birds, planes, and meteorological and astronomical phenomena as 'UFOs.' The current cultural template and active folklore regarding anything strange seen in the sky is automatically to dub it as a 'UFO,' whilst the media automatically associates the acronym with 'aliens' or 'extraterrestrials.' As Jenny Randles has said about the interaction between witness and media, "'Woman sees spaceship' is news, whereas 'Woman sees spaceship, but was probably mistaken' is not."[29]

It does not matter if the witness is from a background of trained observation, such as a policeman or pilot. There is no such individual as the totally credible witness. *Everyone* is equally susceptible to misperception in that magic moment when they see something unknown. If a witness already has a belief system to fit a 'UFO' sighting into, or a predisposition to 'believe' in any aspect of the supernatural, the situation is further exacerbated. Tony Dodd said of his increasing sightings in the Cracoe area during the 1970s and '80s that "the extraordinary was becoming the ordinary." The sighting by the police officers at Cracoe was a case of the reverse of this, the moment the ordinary becomes the extraordinary. This moment, the transition from 'ordinary consciousness' to 'UFO consciousness,' is at the very core of the UFO mystery.

Unfortunately, UFO buffs fail to learn from the significant lessons taught by cases such as Cracoe and others like them. They soon forget the wild claims, the 'scientific' analyses, the far-out theories. Instead of applying the principles learned to other cases, they move on to the next 'unexplained' case in the hope it will be the 'big one.'

Eventually, in the face of overwhelming evidence, YUFOS underwent a paradigm change. In 1987, by which time no one now believed in the 'structured craft' interpretation of the Cracoe photographs, Tony Dodd wrote a piece for *Quest* magazine in which he re-framed the Cracoe photographs as being a rock reflection. There was no mention of the dramatic struggles between rival viewpoints about the nature of evidence which had been necessary to get to this point. By 1997, Graham and Mark Birdsall, now editors and owners of the hugely successful *UFO Magazine (UK)* and *The Unopened Files*, felt able to use the Cracoe case in a round-up of sightings for the fortieth anniversary of Kenneth Arnold's sighting. This time the photograph was captioned: "Lights photographed on Cracoe Fell, near Skipton, North Yorkshire on 16 March, 1981. The 'UFO' is in fact simply sun rays striking the rock surface."[30]

The extraordinary had become the ordinary once again.

6

THE ISLE OF LEWIS MYSTERY

David Clarke

"It had all the ingredients of a case for The X Files. *Strange phenomena in the sky observed by a remote community. Defence forces being scrambled in a massive search, then offers of a convenient explanation ..."*
— Michael Horsnell, *The Times*, October 28, 1996.

The group of islands which form the Outer Hebrides provides the perfect setting for a story which contains all the elements of mystery and the imagination which constitute the growing UFO mythology. Backed up against the vast Atlantic Ocean on the extreme north-west of the British Isles, they are a rugged yet beautiful archipelago, a landscape where weather conditions can change rapidly and without warning. The small population of crofters and fishermen who live there are a highly traditional and religious people who do not readily make sensational public claims concerning strange and unexplained phenomena which so fascinate the population of the British mainland.

The afternoon of Saturday, October 26, 1996 was to change all this when a strange aerial phenomenon occurred in the sky over Lewis, the largest and most northerly of the Western Isles. The event was triggered by calls to the coastguard from a number of people at the Butt of Lewis, on the far north-west coast of the island. They, and others nearer the town of Stornoway, reported hearing a loud explosion and seeing a flash, which lit up the late summertime sky. Some observers were emphatic that what they had seen was a small aircraft, while one individual described seeing a dark object spiralling from the sky into the Atlantic Ocean. A trail of dense black smoke some ten miles in length was left in its wake, 15 to 20 miles out to the west of the Butt of Lewis.

The HM Coastguard log of the incident notes how the first reports were received at precisely 5.05 pm when "a member of the public stated he had seen what appeared to be an object falling into the sea west of the Butt of Lewis.

Prior to this, a flash of light had been seen followed by a loud bang. There remained a trail of smoke in the sky."[1]

One of the witnesses was Norman MacDonald, 57, a joinery contractor who lives in the village of Port of Ness. He first thought the object he saw in the sky was a firework.

"I noticed this flash out of the corner of my eye," he continued. "I stopped and had a look, and there was this streak of black smoke in the sky. I was really surprised to see such a thing. It looked like a small plane with a damaged wing, and I just stood there watching it in complete surprise. I saw three flashes in total and heard a further two bangs. I rushed into the local shop and took the staff and customers out. They also saw the dense smoke spiral."[2]

Another witness, Fred Simmons, was even more emphatic in his statement about what he saw in the afternoon sky. He remains convinced it was a conventional airplane in trouble and about to crash.

"I was out exercising my dog when I saw the plane which was visible in the sky coming over," he said. "There was a lot of smoke coming from the rear end as it headed towards the sea, and it was declining rapidly. I knew something was wrong, very wrong, and I could hear a noise in the distance like thunder. I was concerned; I wasn't frightened because I had never seen anything like that before."[3]

Coastguard staff who quizzed Mr MacDonald and other witnesses at the scene immediately realised how potentially serious these reports were, for the apparent flightpath of the mystery crashing object lay directly beneath one of the major trans-Atlantic corridors for aircraft flying from Europe to North America. The descriptions given by witnesses all tallied with the initial descriptions of flashes and bangs whilst another said he heard "what he described sounded like aircraft coming down." Between 6 and 7 pm that night, "numerous other people" called in with reports of smoke and planes in the area. By 5.17 pm, concern was growing "with the initial report having an aeronautical theme and with a Joint Military Command exercise due to commence shortly. ARCCK was contacted regarding military activity in the area and also to check with the Navy."[4]

As a result, the coastguard launched one of the largest and most expensive air and sea search-and-rescue operations ever seen in the Outer Hebrides, which has been estimated to have cost the tax-payer a total of £200,000. By 5.30 pm, a coastguard rescue helicopter had been scrambled to search "for a possible downed aircraft," Stornoway lifeboat was launched, and a call sign was broadcast to all shipping in the area asking them to keep a look-out for debris or oil slicks floating on the surface of the ocean. However, all vessels contacted said no distress calls had been heard on either the 121 or 243 MHz bands where they would have been expected if an air disaster had occurred.

Senior Stornoway coastguard officer Ian Lindsey explained how the emergency operation progressed from the point at which the first distress calls were received at the Stornoway HQ:

"On receiving these reports, we went into the full emergency procedures of sending out air rescue helicopters, lifeboats, etc., as the area that was indicated to us was below a direct oceanic route for aircraft on trans-Atlantic flights. We really were expecting the worst when we set off from Stornoway that day, but absolutely nothing materialised at the end of the day. We were not sure what was going on, and one of the first things we did was check with Air Traffic Control centres. They did a headcount of all the aircraft which had overflown the area and were able to confirm that everyone was down safely.

"We then proceeded to check local airfields, including the one at Stornoway, microlites, that sort of thing, and managed to rule those out. The Regional Control Centre then linked up with the NASA's tracking station in Virginia to see if there was any space debris that was due to come in, and found they couldn't confirm anything either. RAF Kinloss contacted the satellite tracking station [RAF Fylingdales, North Yorkshire] and ruled that out quite early in the search.

"When the BBC filmed a short documentary on the search for their *Mysteries* series, they interviewed a number of people who said they saw something that day. People did contact us to say they had seen an explosion in the sky; on the day we got about three reports from people. They were good eyewitness reports of people hearing a bang, but only one person saw a ball of smoke in the sky and something spiralling down into the water, a black object of some sort."[5]

The Air-Sea Search Operation

During the search operation, the coastguard co-ordinated the surface search and RAF Kinloss took over control of the helicopters and RAF aircraft. In addition, the police and coastguard mounted shore patrols around the coastline of the island. The air-sea search covered 1,000 square miles of the Atlantic off the north-west coast of Lewis, which was thoroughly and systematically searched using the very latest equipment. Initially, a Stornoway-based coastguard vessel put to sea, but this was soon joined by a flotilla of other small boats, including a French fishing vessel the *Simon Keghian*, and the *Portsalvo*, a Stornoway-based tug. They were accompanied from the air by a coastguard helicopter from Stornoway, which flew two four-and-a-half-hour sorties, and a Sea King from RAF Lossiemouth, which helped conduct an intensive search. Both were fitted with the very latest sophisticated electronic search-and-rescue equipment.

An RAF Nimrod aircraft from Kinloss, call sign Rescue II, joined the operation at 7 pm and continued searching until midnight, when the operation was called off because of poor weather conditions. The Nimrod flew a second sortie at 7 am the following morning, when the crew were able to make use of its detection equipment, capable of picking up tiny objects floating in the ocean from thousands of feet above. The crew sent half-hourly SITREPS (situation reports) to RAF Kinloss. One of these, overheard by a UFO enthusiast on his scanning radio at 8 am on October 27, was reproduced in several UFO magazines. It read, "Confirm ... six feet long and three feet in diameter."[6] This and further messages intercepted later that week have been interpreted by some UFOlogists as evidence that a low-key, secret military operation was underway to retrieve an object which they presumed had crashed into the sea off Lewis. From this point of view, the 'official' search was merely a public relations exercise to cover what was really going on.

UFO researcher Nick Redfern concluded that contrary to what was admitted officially, some form of object was, after all, located during the huge search-and-rescue operation, "but quite what that something has never been made public."[7]

The truth, however, was more mundane than mysterious. In this case, the object described by the Nimrod crew, overheard by a radio scanner, as being "six feet long and three feet in diameter" had a very down-to-earth explanation. It was, according to coastguard officer Lewis, simply a plastic oil drum – the one single object found floating in the ocean search zone during the entirety of the operation. "The Nimrod found it from 2–3,000 feet, which just shows how technical their equipment is and how efficiently it works," he said.[8]

These rumours and speculations were encouraged by a further coincidence. A pre-planned Joint Services NATO training exercise began off the coast of Lewis just two days after the search for the unknown crashing object. It involved seven submarines, 80 military aircraft and 32 ships, including the US frigate *Aubrey Fitch*. This coincidence only added to the stories that were already circulating which suggested that the incident had been caused by a

UFO or military blunder, perhaps involving an aircraft shot down by a stray missile. A further 'community rumour,' heard by a Scottish National Party member, was to the effect that a naval frigate taking part in this exercise had

POLICE PURSUIT

In October 1973, the eastern and southern states of the USA were experiencing a major flap. There had already been several close encounters and claims of alien contact. Late on the night of October 17, Jeff Greenhaw, police chief in the small town of Falkville, Alabama, received an anonymous call. The woman claimed that a UFO was hovering over a nearby intersection. Noting her distress, Greenhaw set off in his patrol car, but on arrival there was no UFO and no phone caller – just a dark edge of town location and a dirt track leading away.

Suddenly, a figure appeared ahead. It seemed to be a human-sized being dressed in a silver suit. The clothing reflected the headlights of Greenhaw's cruiser and stood there, staring silently at him.

Reacting quickly, Greenhaw grabbed his camera and snapped some quick photos. This act seemed to shock the entity, which turned and fled down the dirt track. Police chief Greenhaw clambered back into his car and set off in pursuit. Riding the bumpy track in the dark proved difficult and the silver-garbed figure vanished into the night, leaving Jeff with a tall story to report ... except, of course, that he now had pictures!

Is it real?
● There is no doubt this incident occurred. The police call was logged, but never traced, and the photographs serve as proof. These have never been disputed and depict a human figure wrapped in what resembles kitchen foil.
● Greenhaw had an excellent reputation in the town. Nobody considers his story to be a hoax.
● There had been previous sightings of entities wearing silver suits like that seen at Falkville.

Is it solved?
● Nothing in the photographs even suggests an alien origin. The figure could easily be a human dressed in a suit. No UFO was ever witnessed.
● A similar incident occurred a few days earlier and was reported by the Press. In that case, a hoaxer was caught scaring motorists dressed in silver foil, standing by the highway and pretending to be an alien.

● Greenhaw appeared to be the subject of a vendetta. Some weeks later, his home was attacked by a firebomb and badly damaged. He was ultimately forced out of town.

Conclusion
It seems probable that this incident resulted from a prank played upon the police chief, possibly by criminal elements who wanted him out of office. The Press story about the hoax on the highway and the on-going UFO wave no doubt provided immediate inspiration.

The hoax call drew the officer to the location, but the taking of photographs was possibly not predicted, causing the culprit to flee. It was no doubt anticipated that there would be serious effects on the credibility of the police chief, perhaps suggesting revenge was a motive.

When his honesty and the pictures ensured that Greenhaw's story was believed, further desperate measures may have been required to complete the job, resulting in the fire.

Further reading
Alien Contact. pp.72–5, Randles, J. Collins & Brown, 1997.

been seen by the crew of a fishing boat off the west coast of Lewis "uplifting wreckage from the area – as if it was involved in a recovery operation."[9] These are precisely the ingredients necessary to seed the foundations of a 'UFO crash retrieval' rumour.

Much was made by UFOlogists of the rapid response by the RAF, who took the co-ordination of the search swiftly out of the hands of the HM Coastguard soon after the first reports were received. The sinister connotations of this development were increased by an 'off the record' quote from a coastguard insider, who said that "the buzz was that a missile had gone seriously wrong and was brought down by the military. The intensive search was not so much to discover wreckage, but to ensure nothing was found."[10]

A more prosaic explanation was put forward by coastguard spokesman Phil Smith. He said there was nothing strange to be read into the fact that the search operation was co-ordinated by the RAF. "The coastguard are responsible for co-ordinating civil maritime search and rescue and the Rescue Co-ordination Centre at RAF Kinloss is responsible for aeronautical rescue co-ordination," he said. "As it was initially considered aeronautical, it was not uncommon for them to take over the organisation of the search."[11] Similarly, rumours which suggested the involvement of a stray missile were quickly ruled out by the MoD. Colonel Andy Pedders at the Royal Artillery's Firing Range on the Isle of Benbecula told a local newspaper: "I can categorically assure you we were not firing and had not launched any targets on Saturday or Sunday. It [the incident] has absolutely nothing to do with us."[12]

Chris Murray, one of the coastguard helicopter crew who took part in the search, said the reports that reached them from members of the public were "very alien and very strange," and that when setting off from their Stornoway base they did not know what they were searching for. "We expected to find some sort of debris in the water, although we did not know at that time exactly what had crashed, if anything had crashed into the sea at all," he said. "We carried out a systematic search of the area, but we found nothing at all."[13]

Despite the unusual nature of these events, it was not the first time that a report of this kind had occurred in the sky above the Hebrides and triggered a widescale search of the ocean. In February 1961, for instance, an air-sea search was launched after the skipper of an Aberdeen-owned trawler saw an object crash into the sea off the east coast of Lewis.

The skipper contacted Stornoway coastguard on his radio to report seeing what he thought was an aircraft appear suddenly out of a heavy cloud while his crew were fishing in bad weather. The object flew low, weaving from side to side before falling into the sea near the entrance to Loch Seaforth, causing a splash which was followed by "a column of black smoke rising up from the position." The skipper said: "It was not my imagination. Several other members of the crew saw it, too." Subsequently, Stornoway coastguard broadcast a message asking all ships in the vicinity to keep a look-out, and two vessels spent three hours searching the waters where the object was seen to fall. The search was eventually abandoned after no aircraft were reported missing. An Air Ministry spokesman was quoted as saying the report was being treated as "a false alarm."[14]

The similarities between the two incidents suggest both could have a similar, natural explanation in the realms of meteorology and astronomy. UFOl-ogists tend to seize upon reports such as these as evidence of military cover-ups and retrievals of alien craft or secret technology, but the truth is often of a more mundane nature. The involvement of UFOs in the events of 1996 was first suggested by newspaper stories which paraphrased a spokesman from RAF Kinloss, who confirmed that reports of "three UFOs seen flying in formation" in a northerly direction over the Scottish borders had been received on the day of the explosion.[15] If correct, this observation is most likely to have been one of the same group of sightings which triggered the original coastguard search of the sea.

Three weeks before, a passenger on a trans-Atlantic flight from Vancouver passing near Stornoway reported seeing what he described as an exploding fire-ball in the sky. A pilot friend phoned the *Stornoway Gazette* to report the sighting and was surprised to hear the observation had occurred within weeks of the mystery 'explosion' over the Butt of Lewis.

He said: "My friend saw a light bordering on the horizon and thought that it was an aircraft or a ship exploding. At the time, the pilot was through speaking to the passengers and he was the only one that saw it. He never reported it, but is a level-headed man and wouldn't make something like that up."[16]

In the case of the Lewis incident, Ian Lindsey, Stornoway coastguard's senior watch officer, was able to rule out UFO activity as having played a part in the events of October 26. He said: "I don't want to be unkind to them, but more than one and a half years later people have come out of the woodwork saying they saw this or that who had not reported seeing anything to us at the time. Other people said they saw something coming out of the west, but these 'UFO' sightings have only appeared 18 months after the incident happened and were not floated about on the day we were out searching. The stories only started to appear afterwards, when articles began appearing in UFO magazines. There was no UFO activity reported to us on the day of the incident. It was always treated as a straightforward search and rescue operation."[17]

Military cover-up?

Rumours which suggested the Lewis incident had been caused by a secret military operation soon prompted local politicians to ask questions of the Defence Ministry in the House of Commons. At one stage, the Scottish National Party expressed concern about the incident and asked the government if it was aware of any "intrusion in UK airspace by foreign aircraft" and sought an assurance that "at no time was any danger posed to local fishermen."[18] But they were forestalled by the Labour MP for the Western Isles, Calum McDonald, who, in a written question, asked the then Defence Minister Michael Portillo if there was "any military involvement" in the search for the unidentified object seen crashing into the sea, and if any liaison had taken place with the US authorities in the subsequent search operation.

This was prompted by claims that the crash could have been caused by the malfunction or crash-landing of a super-secret Stealth aircraft, which many people connected with the NATO exercise which began two days later off the Western Isles. However, all enquiries with the RAF and the MoD met with denials that the explosion was linked with any form of military operation. Indeed, RAF Kinloss said they had specifically checked if a Stealth aircraft was flying that day through military channels and had been able to rule this possibility out very early in the operation.[19]

One year later, on October 14, 1997, Lord Gilbert replied in parliament saying that initial reports of an explosion initially attributed to a mid-air collision north of the Butt of Lewis resulted in an extensive search, but this was abandoned "after it became clear no aircraft had been reported overdue." He added that the HQ US 3rd Air Force were also approached by the rescue authorities, but had confirmed "that there had been no US military activity in the area."[20]

The speculation connecting the incident to secret aircraft actually dated back to 1992, when several newspaper reports suggested that the US Air Force had been testing its hypersonic spyplane, code-named Aurora, in the more remote parts of northern Scotland.[21] The Aurora is believed to have had a budget of £10 million from the 'Black Projects' programme, and been flying since the mid-

1980s, even though the project is not recognised in any official US government documents or records. Defence experts were quoted as saying that the plane, which is said to have a top speed of 5,300 mph (eight times the speed of sound), "has been buzzing around the Western Isles for some time, rattling windows in crofts."[22] These Press reports followed accounts from credible witnesses, including a police chief inspector in the Hebrides, of "mysterious rumblings" and sightings of "a mysterious white object shooting across the sky."

Although there is no clear description of the Aurora, it is thought to have a sleek, swept-triangle body without conventional wings. One sighting which has been associated with the craft was that of oil rig worker Chris Gibson, a former member of the Royal Observer Corps. He spotted a "triangle" object in the sky behind a Boeing KC-135 tanker aircraft, flying in formation with two US F-111 bombers at low-level over the sea off the East Anglian coast in August 1989.[23]

The Aurora is, in fact, just one of a number of prototype aircraft, including Unmanned Aerial Vehicles (UAVs), which are suspected of having been tested over the sea off the western coast of Britain. In 1998, *Jane's Defence Weekly* spokesman Paul Beaver went on record to say that a malfunctioning UAV was the most likely contender for having sparked the search off the Butt of Lewis.[24] Indeed, just such a tiny unmanned aircraft, developed by a private company with the backing of the US Navy, was successfully flown across the Atlantic Ocean and guided down to the British Army missile range landing strip on the island of Benbecula in November 1998.[25] However, there is no direct evidence that such an aircraft was in the Lewis area on the afternoon of October 26, 1996, and the limited military involvement in the subsequent search does not equate with the reaction one would expect if a top-secret aircraft had indeed been feared lost.

Similarly, attempts to connect the incident with the Joint Services Exercise in the sea off Lewis which began shortly afterwards fail to convince. This kind of exercise takes place annually in the same area of the North Atlantic, so much so that coastguard stations regard it as a regular calendar event. It was just a coincidence that this particular exercise commenced shortly after the mystery incident, but in the world of UFOlogy coincidences are always highly significant and are regularly regarded as 'proof' of military involvement and high-level cover-ups of retrievals involving 'alien' craft.

Coastguard spokesmen pointed out that at the time of the incident there were no military vessels in that area because if there had been any, they were duty-bound to assist in the search operation as it was put out on all radio frequencies as a May Day.

Ian Lindsey said: "They would have been duty-bound to stop what they were doing and come to help. There was no military activity called up at all, other than the Nimrod which RAF Kinloss sent to help us. I don't believe the incident was caused by the military missile, or by some secret aircraft test. At the end of the day, the truth always emerges. Someone talks or they get found out. The Royal Artillery missile range at Benbecula provides daily weather reports to the coastguard stations, and I can confirm they were not even open that day, having checked the records.

"Then there were the reports of people who said they saw ships dredging the seabed off Lewis, but these coincided with routine exercises which had nothing to do with this incident. Because this is such a remote area, you can read whatever dark and desolate things you want into this little incident, but unless someone comes up with some physical evidence of some sort you can't simply accept these claims."[26]

Meteorites and meteoroids

As we have seen, crashed UFOs, secret aircraft and conspiracy theories found ideal breeding grounds in the mystery which surrounded the Lewis incident. Often drowned out by sensational claims such as these are the less exciting down-to-earth explanations. A far more likely contender as an answer for the Lewis incident was a meteorite or piece of man-made space junk burning up on re-entry into the earth's atmosphere, causing a bang and flash and giving the impression that an aircraft had exploded and crashed into the sea. Debris such as this can appear at night as flashing lights or strings of moving lights in the upper atmosphere. They have been known to leave a glow or a trail of smoke in their wake and be accompanied by loud explosions.

Dr Jacqueline Mitton, of the Royal Astronomical Association at Cambridge University, described the characteristics of bolide meteors in the following manner: "They often look like a series of lights in a trail when they are breaking or burning up in the atmosphere, and have been known to cause explosions or sonic bangs. Very bright bolide meteors are not uncommon. I have seen one myself and it even left me puzzled. The one I saw seemed to move very slowly across the sky, and for people who are not familiar with the night sky it would be very easy to see it as a slow-moving object with lights attached."[27]

Scientists from one of Britain's leading research centres for fireball meteors at Armagh Observatory, Northern Ireland, took a keen interest in the Lewis event. Professor Mark Bailey first heard about it via a colleague at the top-secret Sandia National Laboratories in New Mexico, USA, which operates a network of spy satellites. His contact there was interested in the event "to see whether it might confirm a theory he had been developing concerning the spontaneous detonation of methane gas clouds vented upwards from the earth."[28] The interest shown by Sandia, a centre said to be involved in covert US government investigations into UFO phenomena, was seen as "sinister" and highly suggestive of ET activity by many UFOlogists, including conspiracy expert Nick Redfern. He connected Sandia's involvement with the military exercise off the coast of Lewis, which he said "for me ... was too coincidental."[29]

While Sandia may well have a long history of involvement in the UFO issue in the USA, in this case, their interest in the Lewis case was purely for defence and scientific reasons. Colleagues at Sandia, who had heard about the event via the US space tracking station in Virginia, wanted accurate co-ordinates from Professor Bailey for the location of the explosion to enable them to check satellite pictures which may have captured it on film. The US government's spy satellites are continually pointed towards the earth specifically looking for evidence of nuclear explosions and testing. It is important for scientists monitoring these pictures to be able clearly to distinguish between the explosions produced by meteors and other space junk from those caused by missiles and tactical nuclear warheads secretly tested by foreign governments.

Professor Bailey said colleagues at Sandia were "very interested in the flux of small bodies less than ten metres in diameter and the explosions they cause in the atmosphere ... There is a lot of scientific and technical interest from the Defence side in these objects as they need to be able to distinguish these explosions from those caused by small tactical nuclear missiles. They need to know what they look like, and that is the real reason for the Defence interest in events like the Lewis one."[30] Following this request, Professor Bailey contacted colleagues at the Royal Observatory in Edinburgh and also spoke to reporters at the *Stornoway Gazette*, who were able to put him in touch with a number of eye-witnesses to the event. He told us:

"The basic facts, as I understood them, were that eye-witnesses had observed a flash and black vapour trail on the afternoon (about 15.40 GMT) of Saturday, October 26. There was concern that an aircraft had exploded in mid-air or been shot down by artillery from the Benbecula range. This led to scrambling by the coastguards and ambulance crews, and the launch of the Nimrod, helicopters and lifeboats. An alternative view was that the trail and airburst were created by a piece of space debris or a fragment of a comet or asteroid, collectively known as a meteoroid. One or other of these explanations seemed most likely, as no plane had been reported missing and the Minister of Defence reported in parliament that there was no secret military activity in the area at the time. This was later confirmed by other sources, despite the NATO exercise nearby. My view, therefore, remains that the event was most likely a fireball, produced by a moderately large meteoroid as it entered the earth's atmosphere. Any other interpretation would seem to require a very high degree of 'conspiracy' factor, for which I have no evidence."[31]

Coastguard watch officer Ian Lindsey said this was precisely the same conclusion reached by the RAF's Regional Control Centre after consulting space scientists at the completion of the search. He confirmed the end result of the checks was that the incident was officially explained in the official report to the MoD as being caused by "meteor activity." They estimated that the meteorite involved could have weighed one ton and might have been travelling at 12.5 miles (20 kilometres) when it exploded with the force of 20 tons of TNT above the Hebrides.

"That was taken down as the official explanation," said Mr Lindsey. "It was left open, but the bottom line was that it was meteor activity. This explanation was backed up by rumours in the town that someone had found lumps of what they thought were meteor debris in their garden or croft.

"The meteor theory seemed to me and everyone else to be the most reasonable and most likely explanation. I have worked as a coastguard in the Hebrides for some time, and we get a lot of reports from people who have seen green lights flashing across the sky, and have seen lots of meteors ourselves. We even have a list of dates in the coastguard watch when the most spectacular showers are expected; this incident occurred shortly before the Leonids [one of the largest meteor showers] in November."[32]

Bolide meteors are common phenomena and sightings of blazing fireballs crashing to earth have frequently triggered outbreaks of reports from members of the public describing UFOs and "crashing aircraft." A bright fireball meteor has been invoked by science writer Ian Ridpath as the initial trigger for the scare which swept RAF Woodbridge in Suffolk on December 26, 1980, which provoked a flurry of speculation about UFOs and government cover-ups.[33] More recently, spectacular meteor showers witnessed by numerous witnesses in Britain on the evenings of June 11 and July 10, 1998, resulted in a series of calls to the police and emergency services reporting "flying saucers," "distress flares" and "blazing aircraft." The July 10 shower was accompanied by a large explosion in the sky above the Isle of Man, and the appearance of vapour trails in a Z and Y formation in the night sky.[34] Meteor showers and bright bolides can clearly be mistaken for more exotic phenomena even by observers like Dr Mitton who are experienced observers of the night sky. They are often spectacular, unexpected and may be accompanied by loud explosions and other optical effects.

The British Geological Survey do record a number of sonic events on their sensitive seismic microphone equipment which have been attributed to atmospheric explosions caused by bolide meteors and man-made space junk disintegrating within the earth's atmosphere. Unfortunately, due to the paucity of recording stations in the far north-west of Scotland, no such confirmatory signal was detected on the afternoon of the Lewis incident.[35]

However, one instance published in the BGS' annual bulletin *UK Earthquake Monitoring* describes a very similar event recorded in northern Britain and Scotland early in the morning of September 23, 1997. Shortly before 8 am that day, numerous reports reached RAF Kinloss and the media from people in Northumberland, Edinburgh, Glasgow, Inverness and Wick, Grampian, describing "flashing lights in the sky" accompanied by "a loud bang and windows shaking." Data from the BGS' rapid-access seismograph networks in Orkney and Moray were checked and "a signal consistent with an atmospheric origin was recorded on four seismograph stations and two low-frequency microphones."[36] RAF Flying Complaints were contacted, but "could not confirm whether any military aircraft were in the area at the time."

Later it was concluded by the BGS that the lights and bang were caused by an object burning up in the atmosphere above northern Scotland. This was attributed to a meteorite or a fragment of a Russian satellite which re-entered the atmosphere at around the same period. Other sonic booms have been attributed to military aircraft involved in manoeuvres, but these are often recorded over the Atlantic Ocean and North Sea where prototype aircraft like the Aurora have been test-flown by the US Air Force and the British MoD.

It is little wonder that sightings of these spectacular celestial phenomena often result in calls to the emergency services and searches for aircraft which are presumed to have been lost or crashed in remote and inaccessible areas. One excellent example occurred on the evening of December 1, 1997, when South Yorkshire Police scrambled their helicopter to search the Pennine moors following reports of "a blazing object" crashing to earth. On this occasion, John Barker, a resident of a village high up on the moors at the border with West

Yorkshire, called police to report seeing what he thought was an aircraft crashing into the moors.

"I was sitting in the living room of my house facing the moors when I saw what looked like a multitude of coloured lights like an oxy-acetylene torch through the window," he told me. "It had all colours of the rainbow, including orange, yellow, magenta and green, and I could see burning debris dropping from it. It seemed to drop down or land on the moors, and then it just disappeared. I was concerned that it could have been a small aircraft and because of the bad weather they would be in trouble."

The police helicopter searched the area of fells near Crow Edge where Mr Barker saw the lights descend, but found nothing. Later they discovered that police and UFOlogists in Greater Manchester had also received reports on the same night describing "a bright green or yellow light with tails trailing behind" falling from the sky in Oldham, on the western side of the Pennines. Air traffic control at Manchester Airport said they had not picked up any unidentified aircraft on their radar. Subsequently, it was revealed that both groups of witnesses had observed a huge meteor which crashed to earth that night, not in northern England, but thousands of miles away, disappearing into the glaciers on the island of Greenland in the Arctic Ocean.[37]

This evidence clearly demonstrates how easily eye-witnesses can be mistaken in their descriptions of unidentified objects in the sky, particularly at night when even qualified observers have difficulty judging the distance and height of 'UFOs.' Many UFO reports can be explained as misperceptions of natural phenomena and aircraft under unusual atmospheric conditions. Clearly, the Lewis incident is a prime example of this phenomenon at work. In this case, some witnesses remain adamant they saw an object like a small plane before the explosion was heard, yet we know that no earthly flying object was responsible.

Lewis: UFO crash retrieval?

Taking all the available evidence into account, the explosions and smoke reported over the Butt of Lewis which triggered a major air-sea rescue operation was most likely caused by a bolide meteor or space debris of the kind described by Professor Bailey. Even so, many of the observers who witnessed this explosion remain convinced that they saw a 'real' aircraft, and speculation continues that the event was 'covered-up' by the Ministry of Defence. This kind of mistrust has continued due to the lack of any authoritative explanation or statement from the MoD following the incident.[38] Indeed, the void left by such a statement has allowed rumour and speculation to proliferate.

Since 1996, a number of other unconnected events in and around the Western Isles have become associated with the earlier incident, including mysterious distress bleeps which triggered a five-hour search by the coastguard just weeks later, and the discovery of pieces of a military helicopter in the sea off Lewis at the end of 1998.[39] This find consisted of a rotor head with four blades, and part of a gearbox belonging to a twin-engined Westland Lynx, which was dragged up in the nets of a trawler from the depth of 1,000 feet 80 miles west of the Butt of Lewis. Despite the fact that both mysteries were ultimately explicable in down-to-earth terms, they contributed to the evidence for a conspiracy in the minds of those who are looking for evidence of alien involvement.

Norman MacDonald, one of the eye-witnesses to the Lewis event, maintains to this day that "someone is hiding something" about the incident "because there have been so many reports and so many witnesses, and yet no one has admitted that anything out of the ordinary went wrong or even happened that particular day."[40] The continuing mystery has encouraged the UFO industry to incorporate the story into the self-perpetuating mythology of government cover-ups and secret retrievals of mystery objects.

Sociologists Robert Bartholemew and George Howard argue that this type of story constitutes what is known as 'living narrative folklore' which fulfil and satisfy hidden psychological needs among those who promote them.[41] Folklorist Jan Brunvald has defined urban myths as having three key elements – "A strong, basic story-appeal, a foundation in actual belief, and a meaningful message or moral."[42] Crashed UFO and government conspiracy theories easily satisfy all of these criteria.

The influence of belief and the existence of a "strong basic story appeal" was given credence within a week of the Lewis incident by the appearance on the scene of what the Press called "Britain's top UFO hunter," Nick Pope. A civil servant, for a period of three years Pope staffed the Ministry of Defence's public relations department which deals with enquiries about UFOs from the public, a post which led to him receiving the nickname 'Spooky,' an allusion to the fictional FBI agent Fox 'Spooky' Mulder in the popular sci-fi TV series *The X Files*. Pope has since become an authority on the subject. Although his former post attracts lots of sensational media headlines, the 'UFO investigator' role was, in fact, just a part of his duties in a department which handles complaints about low-flying military aircraft.

Pope claims he became convinced about the reality of extraterrestrial UFOs regularly intruding into UK airspace while working for Secretariat Air Staff 2a. After moving to a new post, he has achieved widespread media attention as a result of 'going public' with his beliefs, which were expounded in two popular books on UFOs and aliens which have become best-sellers in the UK and abroad.[43] In 1999, Pope told me he had not launched his own investigation into the events, as was suggested at the time by national newspapers. "I have drawn no firm conclusions, and remain open-minded as to what lay behind this event," he said.[44] But crucially, three years earlier he was quoted in *The Scotsman* and other newspapers as connecting the event with other reports of 'crashing UFOs,' the most famous of which, he said, was the legendary Roswell incident. Pope's suggestion of a UFO connection with this incident was the first step on its integration into the theories of those who claim the UK government has secretly retrieved pieces of crashed alien craft.

One of the new aficionados of the government cover-up/crashed saucer mythology in the UK is writer and researcher Nick Redfern, who is convinced the MoD has evidence of crashed UFOs and even operates its own super-secret SAS-style saucer-recovery team. He has included the Lewis case in the third part of his trilogy of books charting what he believes is the UK government's UFO "covert agenda," and says of it, "I have come to absolutely no conclusion beyond the fact that something very strange occurred, and that 'something' was retrieved."[45]

Redfern noted the event occurred shortly after a flurry of UFO reports from East Anglia, which involved sightings by the police and coastguards, and an unknown object tracked on RAF radar. Shortly after these sightings in early October, 1996, Nick "received a veritable deluge of telephone calls" from contacts reporting the mysterious events off the north-west coast of Scotland. This was too much of a coincidence for a seasoned UFO researcher. Redfern was able to conclude that once again "a wave of countrywide UFO encounters had been followed by the crash (and possible retrieval by the military) of an unknown object."[46]

Former Yorkshire police sergeant turned 'alien investigator' Tony Dodd featured the Lewis mystery in a chapter of his latest ET best-seller, within the context of claims that aliens had established huge secret bases beneath the waters of the North Sea. From Dodd's viewpoint, the descriptions of an object crashing into the sea and a subsequent military search were entirely consistent with reports he had received from trawlermen in Icelandic fishing grounds.

These anonymous contacts claimed to have seen huge UFOs moving in the sea beneath their vessels, in addition to taking off and returning to the sea, sometimes in full view of US naval frigates.[47]

These are precisely the sort of connections and coincidences which make UFOlogy an exciting and glamorous subject, and provide good material for popular and often lucrative books promoting sensational claims. But at the end of the day, the evidence to back up the claims of a UFO retrieval in the Lewis

THE WELSH TRIANGLE

During 1977, West Wales was in the grip of a major UFO flap. It began in February when pupils at Broad Haven Primary School claimed to have watched a landed UFO "as long as a couch" over several hours. UFOs were seen frequently that year and many sightings also involved entities.

The most bizarre happenings took place at Ripperston Farm, where the Coombs family were plagued by lights and UFOs. These events culminated when they saw a tall, silver-suited figure in their garden. Billy Coombs also claimed that cattle had often vanished from his farm, reappearing on neighbouring farmers' land.

Is it real?

● The sheer number of sightings argues for a reality to the case.
● Several of the schoolboy witnesses drew similar sketches of the UFO seen on the ground at Broad Haven school.
● The Coombs' claim their cattle could not have opened locked gates

and passed through the farmyard without them being noticed.

Is it solved?

● None of the children at Broad Haven School thought to approach the 'UFO' and no teachers seemed interested. Would a genuine UFO stay on the ground for several hours within yards of a school? The UFO sketches were not done until three days later when initially disparate reports of size and shape had become homogenised into a saucer-shaped craft. An unusual farm vehicle was reportedly in the area at the time.
● In 1996, Milford Haven businessman Glyn Edwards came forward to reveal that at the height of the flap he had "taken a stroll" around Broad Haven in a silver asbestos-lined boiler suit worn by oil refinery workers.
● Radio 4 presenter Ray Gosling interviewed local farmers and discovered that many thought the Coombs family had

simply left their farm gates open, leading to the cows wandering from field to field.

Conclusion

No physical or photo-graphic evidence for the Welsh Triangle exists. The narrative evidence is too varied and confused to relate to any consistent phenomena. Psycho-social theorists have suggested that uncritical media reports led to the local population being 'UFO sensitised.' This resulted in a contagion of belief in which mundane objects were misperceived as UFOs and hoaxes were perpetrated.

Further reading

The Dyfed Enigma. Pugh, R. and Holiday, F. W. Coronet, 1979.
"The Story Of The Welsh Triangle." pp. 810–813. Hilary Evans. *The Unexplained*. Orbis, 1983.
"The Ripperston Farm Riddle." pp. 834–837. Hilary Evans. *The Unexplained*. Orbis, 1983.
"The Truth About The Welsh Triangle." pp. 874–877. Hilary Evans. *The Unexplained*. Orbis, 1983.

case simply does not exist. What is more, it does not equate with the known and checkable facts of the case. The mystery has all the key elements required to fit the 'crashed saucer' narrative during the last twenty years: an unknown object crashing in a remote area, a military operation with no obvious conclusion, bland denials by government officials and an element of 'mystery' which continues to surround the incident. Now it has been established as one of the tenets of UFO crash-retrieval mythology, the Lewis case will continue to resurface in numerous books, magazines and newspaper stories, which ignore the facts and plump instead for speculation.

The evidence assembled in this chapter certainly supports the view that something 'very strange' did occur over that remote island off the north-west coast of Scotland one autumn night in 1996. However, strange phenomena do not necessarily have an ultimate origin which require the presence of extraterrestrial intelligences. Indeed, in this case all the available evidence suggests the Lewis event was caused by an object from outer space. But in this instance it was in the form of a lump of rock or debris known as a meteoroid. From that one simple but unexpected occurrence, claims have been made up about secret military operations, crashed saucers removed from the sea under massive secrecy and a host of other unlikely scenarios, all of which have contributed to the steadily evolving 'crashed saucer' folklore.

Three years after the event, residents of Lewis remain as puzzled and uncertain about what to believe. But as *Stornoway Gazette* reporter Jen Topping concluded in a conversation about the case, "As far as the *Gazette* staff are aware, this remains a mystery, although the meteor explanation seems the most plausible, at least to a sceptic like myself!"[48]

FLASH, BANG, WALLOP – WOT A PICTURE!

The Alex Birch photograph

DAVID CLARKE AND ANDY ROBERTS

"Twenty-five seconds was the time it lasted. Eight seconds of that was took [sic] up in taking the photograph. That is not a lot of time to ruin a person's life; the people concerned should be exposed, and the public should know."
— Alex Birch, www.ufo-images.ndirect-co.uk.

According to official records, the British Ministry of Defence received 46 UFO reports in 1962. One of those was from a schoolboy, Alex Birch, from Mosborough, then in Derbyshire. His report was somewhat different to the usual 'lights in the sky' account. It contained a photograph, the product of a spring morning encounter between young minds and the unknown. This blurred photograph has haunted Alex for almost 40 years, fading in and out of his life like an unwanted and malevolent spirit.

Its impact on UFOlogy was dramatic, too. UFOlogists became polarised between those who firmly believed Alex had captured the proof they longed for and those who saw a different reality in the photograph. All Alex Birch said he wanted was to know what others thought he had photographed. But the simple act of taking – or faking – a photograph was more complicated than that. The alchemy of exposing film emulsion to daylight for a fraction of a second created a monster, changing Alex's life forever. It led him to discover far more about human nature than aliens could ever have taught him.

In 1962, Alex Birch was a fourteen-year-old schoolboy living with his parents, Alex and Margaret, at Moor Crescent in Mosborough, now a suburb of the city of Sheffield, and attending the nearby Westfield Comprehensive School. It was a grey Sunday morning on March 5 that year and Alex was out in a field behind his home with a pal, David Brownlow, aged twelve. Alex had his one-year-old box Brownie 127 camera and was taking photos of a dog when Stuart Dixon, aged sixteen, emerged from a nearby orchard. One photo was taken of Stuart jumping up into the air from a stone fence post. Then seconds later the pair were both amazed to see a group of five strange flying objects hovering in the sky above the fields.

Alex told the *Derbyshire Times* three months later: "They were not moving and they made no sound. They were vivid, just hanging there. After a second or so, some big white blobs started to come out of them and they were sort of hazy and obscured. I got my camera up and took a shot of them. A second or so later, they disappeared at terrific speed towards Sheffield."[1]

The boys went home and told their parents about the strange sighting, but were not initially taken seriously. Because Alex was short of pocket money, it was nearly three months later before the roll of film in Alex's camera was developed at a local chemist. Even then, the chemist did not make a print of the one negative which was later to rock the world because he thought it was spoiled! Undaunted, Alex showed the negative to his mum, who gave him the money to have it developed.

Alex's English teacher Colin Brook, by all accounts a believer in flying saucers, then heard about the sightings, interviewed the boys separately and said their stories checked out. He stated: "There is no doubt that they mentioned the sighting before the photograph was developed. This seems to discount that they are making up a story after having seen what may have been caused by a fault in the lens or in the emulsion of the film."[2] Mr Brook added, "It is unlikely that they have indulged in trick photography as their equipment is simple and the line of trees clearly visible."

Significantly, after the photo was taken, Alex and his friends did not push the 'flying saucer' angle, but appeared to allow others, including Alex's father and schoolteacher, to promote the case for them. Alex told a reporter from the *Sheffield Telegraph* that "the possibility they might have been flying saucers did not cross my mind at the time."[3] Stuart Dixon stated in 1999 that he never thought much of the photo or the sighting. He was convinced – and remains so – that the 'objects' were really a huge flock of starlings which were frequently seen swarming en route to the warm city centre in the winter months. "It was a strange sighting of starlings I will admit," he said.[4]

Overnight sensation

The publication of the photograph in the *Sheffield Telegraph* and *Yorkshire Post* was the start of the adventure of a lifetime for the three youngsters – and for Alex Birch in particular. The picture was soon flashed around the world, and appeared in newspapers and magazines and on TV, triggering a huge wave of

UFO sightings across the city of Sheffield, and in surrounding areas of Yorkshire late that August. Alex's photo was featured alongside these new observations, which included 10 feet of movie film shot by a Sheffield cutlery worker, Walter Revill, of an oddly-shaped UFO soaring in the night sky over Walkley.[5] Astronomers from the city's university said the brilliant planet Jupiter had probably caused most of the sightings. This was confirmed when Keith Graves, a reporter for the *Sheffield Telegraph*, was called out to Rotherham on the night of August 28 to watch the 'UFO,' which had been visible "for the third time in a week, almost exactly at the same hour on the same day of the week."[6]

By this stage, the photo had been enthusiastically taken on board by the newly-formed BUFORA. Later that year, Alex travelled to London to address the inaugural meeting of BUFORA, and told the assembled members what they wanted to hear. The trio filled in detailed sighting questionnaires. When asked what was the most likely explanation for what he had seen, on his form dated August 12, 1962, Stuart Dixon replied, "I think that they were space objects." Strange starlings indeed! Meanwhile, David Brownlow said on his report form, "I believe they was [sic] flying saucers." Alex simply said "flying saucers."[7]

BUFORA threw themselves into the investigation, although at that stage they believed the photographs had been taken in February as the boys could not remember the exact date. It was clear that BUFORA's idea of an investigation was simply to accept the prevailing thoughts of the era, and conclude that the picture showed alien spaceships in the classic 1950s mould. Analysis of the photo by one John Adams did throw up a number of inconsistencies between the boys' story and the image on the photo.[8] But BUFORA's Alan Watts – a hard-core ETH (extraterrestrial hypothesis – i.e. alien visitors) believer and writer of popular UFO books – concluded in one report on the photo, "If we want the truth, I would say we couldn't do better than take these to be fairly normal Adamski-type 'saucers' and argue it out from there."[9]

Waveney Girvan, an aviation writer who edited *Flying Saucer Review* and a firm believer in flying saucers from outer space, said he believed water was the clue behind ET interest in Sheffield. "If there is life of any sort inside these flying objects, it presumably needs water to sustain it," he told the Sheffield newspapers. "It is likely that they have to collect water from the earth for their return journey, and Sheffield is surrounded by reservoirs."[10]

Despite claims that the photo was subjected to lengthy and detailed analysis, there is little evidence in any of the surviving records that anyone other than the Air Ministry looked at the photo and negative in any form of objective fashion. The photo came to the attention of the Ministry of Defence when Alex Birch's father wrote to them on July 2, 1962, informing them about the case and "awaiting instructions."[11] The MoD wrote back to Mr Birch, asking if they could see the original negative, but he was reluctant to part with it. But although the Air Ministry were clearly interested in the case and an official wrote in an internal memo, "I think that as we have shown interest in the photograph it would be as well if someone could examine the negative," this hardly displayed any sense of urgency or importance. Even when Mr Birch offered to allow ministry experts to examine the negative at his home in Mosborough, they declined, saying it "would not really serve any useful purpose. A proper examination will require the use of photographic and optical equipment."[12]

Off to London

On Monday, August 27, 1962, Alex and his father travelled to Department S6 at the Air Ministry at Whitehall, a trip sponsored by the *Yorkshire Post*. But when the group arrived at the building, the reporter was separated and taken to another room while two Flight Lieutenants, R. H. White and A. Bardsley, questioned the boy and his father for what the ministry said was two hours. An internal account of the meeting stated that "both Mr Birch and his son were prepared to talk about it at length … The atmosphere was quite cordial."[13] In 1998, Alex was to claim he was questioned for up to seven hours, and later that year his father was to assert that his son had been placed under "duress" during the interview.[14]

Today, Alex vividly remembers the visit to London as one of the highlights of that year's whirlwind of publicity. He recalled "walking up steps of Whitehall and meeting a man in a tweed jacket, flannels and a dicky bow. We went down long corridors, into a room were there were some men and a doctor. They took the negative and the camera, and took them apart. They asked me all these questions for so long I got muddled. They told me they were not flying saucers but Russians. They asked if I had seen anything like this before, if anything had happened which I had not told my parents, if I could see through them, had I seen anyone in the area the day before?"[15]

The Air Ministry borrowed the camera and the negatives overnight for inspection and analysis by their own photo experts. In an internal memo dated September 24, 1962, which was released 30 years later, Flight Lieutenant Bardsley writes to a colleague in the ministry, "At the risk of boring you, perhaps a brief outline of these doubts will assist you in deciding what on earth you can write to Mr Birch." The document makes it plain those who looked at the picture suspected it was a hoax, and says, "It is a relatively simple task to reproduce an identical photograph to the one we were shown."[16] Officials eventually concluded that the photo had been caused by "ice particles in the atmosphere", but the Air Ministry clearly had doubts. In a letter to Alex's father dated September 25, Flight Lieutenant White commented: "When you brought the negatives along on August 27 for us to have a look at them, two exposures on the film were missing and you explained that these had been spoiled. It is also a possibility that the photograph of the 'flying objects' is the result of an imperfect exposure."[17]

He concluded, "To sum up, the photograph can be explained in mundane terms and does not mean that so-called 'unidentified flying objects' must have been over Sheffield at the time it was taken." A later internal memorandum cast further doubt on the veracity of the photograph, saying, "… the sequence of exposures on the two strips of negatives we saw do not exactly fit the boys' story."[18]

Subsequently, a print of the Mosborough photo from the original negative was submitted to the US Air Force's official 'X-Files' Project Blue Book. The file on the case, now in the National Archives in Washington, DC, states simply, "Insufficient data for evaluation."[19] But there was one strange sequel to the official 'interest' in this case. In 1972, after Alex went on record saying the photo was a hoax, he had left his childhood home in Mosborough and bought a new house with his pregnant wife Anne in Sheffield. But shortly afterwards, they were visited by two mysterious men, who claimed they were from the Air Force and who told him enigmatically they had looked into the case, saying "We think we have sorted it out."[20]

Confessions

Ten years after the flying saucer photo took the world by storm, Alex Birch phoned a number of national newspapers and confessed the picture was a hoax. Initially, only the *Daily Express* showed an interest, but soon afterwards Alex appeared on a BBC2 news programme and confessed all. He told viewers he had taken the photograph of the background of trees through a sheet of glass on which he painted the five 'flying saucers.'[21] He also told the *Sheffield Star* that his father did not learn of the hoax until October 5, 1972, the day before the TV broadcast.

Twenty-five years later, he revealed his father was devastated by the reve-lation and begged him not to go ahead with the plan. Reporters approached David Brownlow, then working as a fitter in the city. What he said tallied perfectly with Alex's account of how the photo was hoaxed: "It was just a joke that snowballed. We just got a piece of glass and painted five saucer shapes on it. Then we took a picture." And he added, "It was just unbelievable that every-body was taken in by it."[22]

None of the papers quoted the views of the third boy, Stuart Dixon. In 1999, he claimed he told journalists he still stuck by the original claim, and that was why he was not quoted as it did not fit the story that was emerging.[23]

When Alex was interviewed in 1998, he was asked what motivated him to come clean about the hoax in 1972. He replied: "What you have got to under-stand here is this: I have since I was 14 years of age had nothing but torment from my schoolmates, from people, everything. The ridicule gave me anxiety and made me feel inferior to people. It caused problems with girls, who would say I was a nutcase. When I met my wife, I tried to keep her out of it and even moved to Derbyshire to get away from it."[24]

This harassment was not limited to his peers. He also spoke about a long procession of "nutcases" who turned up on his doorsteps, and claims he was contacted by US generals and even the late Earl Mountbatten, a known flying saucer buff who firmly believed in extraterrestrials. According to Alex, the *Daily Express* commissioned an artist to produce a reproduction of the famous photo on a sheet of glass. The 24-year-old salesman was subsequently photographed standing behind the glass, holding his Box Brownie camera with a wide grin on his face. He said: "The hardest thing is getting people to believe us now we've admitted it's a hoax. But it's true."[25]

But if Alex thought he would have problems convincing those who had built their belief systems upon his photo, he was going to have even greater problems persuading sceptics 25 years later that his 'hoax' confession was itself a hoax.

Reality check

UFOs and extraterrestrial visitors were big news in 1962. In the 1990s, with the approach of the millennium and the massive success of the TV series *The X Files* and ET movies, aliens and flying saucers were again back in the news. The year 1997 saw the 50th anniversary of the coining of the phrase 'flying saucers' and a new generation of UFO believers were rediscovering the subject which many abandoned when the mirage of hard evidence had slipped through their hands. The Alex Birch photograph was long forgotten and lost in the myriad of claims and dodgy pictures which had fallen by the wayside. In fact only a few UFOlogists now cited it as an example of how once upon a time their predecessors had been fooled by a child, implying it could never happen today.

But the time was right for Alex, now 50 and a successful businessman and antiques dealer, to re-emerge with an incredible story to beat all previous ones. Professing no interest in the subject of UFOs per se, and even less desire to become involved in the loony world of UFO believers who had dogged his life for the last 30 years, Alex Birch said he simply wanted to reclaim the copyright on his photo, which he now maintained was genuine after all.

Alex – by now a grandfather – said he was spurred back into the limelight by his five-year-old grandson, Thomas Dodsworth. "My little grandson always used to point at the TV when he saw the picture I took and said, 'That's your picture, granddad'."[26] Despite the fame of the story, Alex no longer owned either the original negative or a print of his famous photo, even though it had been used in numerous UFO books, magazines and TV shows screened across the world.

Alex was quoted as saying in a local newspaper: "People think I made a fortune out of the photograph – I have heard estimates of £500,000. In fact, I made practically nothing, but I did become internationally famous. But with the fame came a lot of misery. I faced a lot of ridicule and pressure. I decided to claim it was a fake in the hope it would all go away and the pressure would be taken off me. But it didn't work like that.

"The UFO fraternity refused to believe me. They even called an emergency conference in London and came to the conclusion that my change of story was due to pressure. The reason I've decided to let the real story be known now is because I think it is important the public should know. The Roswell Museum in New Mexico want the old Brownie camera to put on display because they believe the picture was genuine. I still have it and it's become an old friend. I have not made my mind up yet about parting with it. But it seems that one black and white picture taken in a garden all those years ago will be having an impact on our lives for some years to come."[27]

The new claims caused considerable consternation among UFO researchers, many of whom were openly hostile to the case, and felt that Alex – a self-confessed hoaxer – could not be trusted in the light of his earlier admissions.

When contact was made with the other two 'witnesses' involved in the 1962 sighting, this unease was initially upheld.

Supporting witnesses

David Brownlow, now 48 years old, still lives in the Mosborough area, and was clearly surprised when confronted with Alex's fresh claims. He said: "It was a hoax. Alex has always run with it more than I have. It was painted on glass. We were just messing around in Alex's dad's greenhouse when we had the idea to do it. We were all into *Quatermass* and *War of the Worlds* at the time. It was Alex's idea to take the photo, but then his dad and a teacher at the school got hold of it and we all got swept along with the hoax, which just snowballed. It was an incredible experience and we had our ten minutes of fame at the time. I just want to forget about it now."[28]

Stuart Dixon, the third 'witness' who was 16 at the time and is now 53, was also contacted in December, 1998, and made the following comments by phone: "It was a fake, and no matter what Alex says that's what it was ... We had a teacher at school who was a UFO freak. I said, 'Let's fake a photo of one,' and so we did. It was a perfect picture. This teacher fell for it straight away. Next thing it was in the papers and on TV all around the world. We just painted them on a pane of glass. If you look at the original negative, you could see the British Oak pub in the background, the chimney stack is at a slant and you can actually see the edge of the pane of glass that we painted the UFOs on to. The more people believed in it, the more it took off and mushroomed. We all agreed to stick together and stay with the story, and that's what we did for ten years."[29]

Several months later, when Alex's fresh claims appeared in the *Yorkshire Post* and *Sheffield Star*, Stuart phoned his old friend and the pair arranged to meet behind the British Oak pub where the sighting had occurred. It was the first time they had seen each other since the early 1980s. Stuart went along despite opposition from family members, who feared association with UFOs could bring further ridicule.

Shortly after this, Stuart contacted David Clarke at the newspaper where he worked and said he wanted to tell his side of the story. At the meeting, Stuart went back on his earlier claims made on the phone several months before, saying, "I would be surprised if I had said that." He maintained he had simply reiterated on the phone the standard 'It was a hoax' line, which he claimed he prepared for anyone who asked him about the photo.[30] This contradicted his stance that he, alone out of the three witnesses, was the only one who had stuck by the original story over nearly 40 years.

Deeper issues

The story of the photograph taken by Alex Birch in 1962 is just the tip of the iceberg in a complex series of experiences which place his claims in a completely new context. Dismissing the whole saga as a hoax which got out of hand is easy, but ignores the important lessons the story can teach us about the complexities of human nature, the mysteries of the psyche and the creation of the UFO myth from very human origins.

For it is clear that Alex, like many others who have made UFO and paranormal claims, is a 'repeater' witness who has experienced a series of strange phenomena throughout his life. Several of the newspaper stories in 1962 referred to other UFO sightings, claimed by both Alex and his parents. Indeed, in a letter to BUFORA in 1963, Alex's father wrote, "… I am anxious to learn the truth about these things that are seen in our skies, of which I have seen two these last two months."[31]

Further experiences were revealed in a series of in-depth interviews conducted with both Alex, his wife Anne and their son Adrian in 1998.[32] As he said in one interview: "We're not daft or eccentric people. All our life we have just accepted and thought it was part of ordinary life, I suppose … It does happen and I always accepted that as a child."

Alex was the eldest of five children born into a working-class family who lived in a modest house on the outskirts of Sheffield. Like many other families in the area, his parents were deeply superstitious, and Alex was told his grandmother was a Romany gypsy and had psychic powers. Alex displays similar psychic propensities, including violent headaches and illness which reach such levels that he is incapacitated during the period preceding a thunderstorm. In 1972, he suffered a classic event – being struck by lightning – which researcher Albert Budden has associated with other UFO percipients. Alex experienced a direct strike while battling to stop rainwater flooding into his home in Nottinghamshire, a devastating episode from which he narrowly escaped with his life, and left him with "an incredible suntan" for several weeks afterwards.

Alex commented: "My wife says I attract these phenomena which happen everywhere I go, and is ever present with me, sometimes with long breaks of no occurrences, then blocks of intense sporadic incidents. I always feel ill afterwards."[33]

Indeed, one of his earliest recollections, from the age of five, was seeing lights dancing around the bedroom at night, bouncing off his hands and forehead. This kind of experience is among the strongest but most frequent features found by Jenny Randles in her study of 'star children,' people who have undergone repeated encounters with aliens and UFOs since their youth.[34] Alex says of this: "You could probably say it was just part of childhood and dreams and imagination, yet it was more than that. These lights were always about the size of a ping-pong ball, darting about the room. They were always a bright fluorescent colour and a green opaque colour."

A brilliant heavenly light was also central to a near-death experience which befell the young Alex at the age of six or seven, and which he interpreted as a "religious experience." It was then that he caught scarlet fever and subsequently diphtheria, and spent weeks on a danger list in a confinement hospital at Lodge Moor. While lying semi-conscious in an upstairs ward with the windows open, he saw a light, approximately the size of a light bulb, appear in the sky. It became larger and larger, and, although bright, did not hurt the eyes. Alex was transfixed by the light, which gave out "beautiful golden rays." These seemed to enter his body. Then, he said, a figure emerged from the light and smiled at him. Without saying a word, it just stood and stared for what seemed to Alex to be hours and hours before the strange experience ended. A week afterwards, Alex was taken off the danger list and made a full recovery. One of the memories he had of the event was telling his father and uncle, who visited the ward, "Dad, I saw Jesus today."

These childhood experiences were just the preliminary to 30 years of strange events involving UFOs, poltergeist-type phenomena, weird, disembodied sounds and voices, and bizarre happenings which have been witnessed not just by Alex, but by other members of his family, friends and neighbours. Three years before his 1962 UFO claim, Alex says he saw a "dumb-bell"-shaped UFO emerging from the ground in a field near his Mosborough home, and moving off, lighting up the ploughed earth beneath it. He went to examine the field afterwards, but found nothing.

Haunted man

The house on Moor Crescent where Alex spent his childhood was, he claimed, persistently haunted, with doors opening and closing by themselves, strange voices on the stairs, and household objects being thrown and smashed. These problems became so bad that at one stage the local vicar was called to bless the house. Alex's mother died from cancer at the age of 46 in 1968. Shortly afterwards, Alex, now a salesman, moved to set up home with his pregnant wife in a terraced house in Jamaica Street, Pitsmoor. The couple had two children, Adrian and Samantha. Here the hauntings continued, along with continual jibes from workmates, which he claims caused him to suffer extreme stress to the extent that he was put on medication by his GP, and eventually led to his decision in 1972 to go public with the 'hoax' explanation of the UFO photo. Alex's father died shortly after the revelation. "It really crippled him," said Alex. "He pleaded with me not to, but I said: 'I can't carry on. I've got a wife and a baby coming on'."

Strange lights in the sky, poltergeists and other phenomena have followed Alex and his family through every subsequent move, from Pitsmoor back to Mosborough, then to Woodhouse and finally to a village in Nottinghamshire. One of the most frightening of the experiences is that of a horrendous 'howling'

THE ROSSENDALE ANOMALY

Witnesses across the Rossendale Valley of Lancashire in 1979 and 1980 were regularly describing a strange object. Said to be an orange light, it moved silently through the sky heading south towards Manchester and provoked much media attention. The area was dubbed 'UFO Alley' as a consequence.

North West UFO group MUFORA (Manchester UFO Research Association) noted a curious pattern. Many of the sightings occurred between 1 and 2 am, and over 75 per cent were in the early hours of a Saturday morning. On Friday night/Saturday morning November 14/15, 1980, Norman Collinson, a police detective inspector, went onto the moors near Bacup.

A member of MUFORA, Norman checked the pattern discovered by the group.

Suddenly, an orange mass of light swooped out from low cloud, causing him great shock. It flew only 200 feet above his head, and appeared to be about to crash into the Pennines. Making a quiet whistling noise, it was surrounded by a line of white lights or windows. The object 'wiggled' slightly, then seemed to regain control, climbed skyward, and with a faint roar disappeared towards Manchester. The police officer had been the latest witness to see what was being labelled "The Rossendale Anomaly."

Is it real?
● Norman Collinson is an impeccable witness, who

went on to a high-flying career. He described precisely what he saw.
● There were dozens of witnesses across this area of east Lancashire who saw more or less the same UFO. Several others reported it to swoop down silently in this way, then roar into life.
● Checks with local air traffic control for 1.25 am on November 15, 1980, revealed no air traffic on radar.

Is it solved?
MUFORA conducted an experiment. They located groups of skywatchers at the appropriate time and day of the week at hilltop positions across Rossendale. Air band radio receivers recorded the call signs of all air traffic through the skies of Lancashire. The 'wiggling'

138

noise – described as being similar to a woman screaming – which plagued the young family after their move to a house on Greenwood Lane, Woodhouse, in the 1980s.[35]

The noise was first heard on Hallowe'en night, leaving the couple paralysed with fear in their bed. It was so loud "you could actually feel the walls vibrate." Despite lasting almost half a minute, subsequent enquiries with next door neighbours found no one had heard anything. But over the next few months, the couple's young daughter heard the same terrible howling while she was alone in the house and was left traumatised. Subsequently, Alex, Anne and their son Adrian have all experienced the noise on up to seven separate occasions as well as other disembodied voices both in and around the house, and at their present home.

Rather than being human or animal in origin, the strange howling appeared to be produced by a vibration or humming, which seems to be electromagnetic

---

light was seen. Meanwhile, member Peter Warrington was in the air traffic control centre at Manchester Airport logging radar traffic.

In the early hours of November 22, 1980 – precisely a week after his sighting – Norman Collinson returned to Manchester Airport to check out the 'culprit' identified by the experiment. He was given permission to board the flight deck and talk to the crew of the suspected cause of the sightings.

The investigation revealed that the flight was a cargo aircraft on a routine trip from Scotland to Manchester. It was operated by an airline whose colours were red. The crew confirmed that every Saturday morning they 'glided' over north Lancashire, switching off cockpit lights, but leaving on

tail fin illumination that caused their logo to glow. This saved fuel and gave them a "superb view of the sky."

Once approaching Rossendale, on final approach into Manchester but below radar coverage, they throttled up the engines and landed normally. The practice was illegal, but widespread, and never engaged in during the day or with passenger flights. Many UFO sightings might be triggered this way.

Conclusion

This UFO was positively tracked to source because UFOlogists suspected an answer, used skill to track it down and were lucky. Having a senior police officer as a member made facilities available to help in the search that would

normally be denied.

The investigation did not solve every sighting in Rossendale, but did uncover a type of IFO that may be creating sightings all over the world. The airline, a major carrier that still flies, has not been here identified, to avoid embarrassment. In November 1998, the hit ITV drama *Heartbeat* built a fictional episode around this unique investigation, giving no credit whatsoever to MUFORA or UFOlogy!

Further reading
The Pennine UFO Mystery. Randles, J. Granada Books, London, 1983.

in origin. A similar phenomenon may have been at work in a baffling event which disturbed the family and neighbours at their Woodhouse home one night in 1984, causing them considerable concern. They said they were woken in the middle of the night by a deep vibration or humming noise that at first appeared to be an electrical fault, but gradually became louder and louder. The couple woke up the family because they seriously thought a major electric supply cable had broken directly beneath the house and were worried it could catch fire at any time. But Alex was dumbfounded when he opened the back door and found the source of the hum seemed to be the raised concrete patio behind the house, "which was red hot and part of the concrete was warm." At the same, time neighbours were emerging from their houses. Car alarms sounded across the street and could not be silenced.

"Then, all of a sudden, the whole lot just stopped just like that," said Alex. "You could hear bonnets shutting and doors closing, and a neighbour said, 'I'm going back to bed.' But we didn't go back to bed, and it was a star-lit night. I turned to my wife and said, 'What's going on?' All of sudden, we saw a bright light come across the fields out of the distance. It was really bright. It got bigger and bigger until it stopped above us standing outside on the patio, then turned and shot off at an angle. And then another one came to the same place, stopped right above us and then shot off again just like the first. We just could not believe what was happening. Everyone had gone back to bed, and the vibration had stopped beneath the patio. It was just a quiet, star-lit night."[36]

These repeated experiences have left Alex open-minded about the origin of the phenomena, which he claims to have experienced over his lifetime, many of which are as puzzling as the photograph he took in Mosborough. Alex initially believed the objects saw in 1962 were 'flying saucers,' despite his later claims about the photo being a hoax. But in 1998 Alex revealed to us his belief that the objects were "... in all probability caused by the same mysterious electrical phenomena that may exist within the earth and the atmosphere of our planet. The fact that one person seems to see these objects more than other people may be that he or she emits a stronger form of electrical magnetism. We are all a chemical and electrical life form. It is probable that I may be a prime example."[37]

This selection of stories from Alex's family's recollections merely skims the surface of the bizarre experiences he tells in a very sincere and forthright manner. As there are other witnesses to some of these events, not all of them direct family members, it seems unlikely that they are all inventions of one man's imagination. And yet if we accept that only a fragment of these stories are 'true,' we have to consider how that affects our interpretation of Alex's famous UFO sighting of 1962, and the photograph associated with it.

Fantasy prone

It is an inescapable conclusion that Alex appears to be the focus of a variety of 'supernatural phenomena,' which appear to be largely electromagnetic in origin. He is not alone in this category, for UFO researcher Albert Budden has interviewed dozens of similar subjects who have experienced repeated UFO and related experiences throughout their lives, which he has linked to "major electrical events," such as electrocution or being struck by lightning. In addition, many other UFO and paranormal percipients have suffered out-of-body or near-death experiences in their childhood, which appear to have set the stage for more advanced phenomena in their later lives.[38]

Meanwhile, psychologists Wilson and Barber, and sociologists Bartholemew and Howard, have isolated what they describe as a category of people who exhibit symptoms of what they classify as "the fantasy prone personality." These are predominantly normal, healthy individuals who are "prone to experiencing exceptionally vivid and involved fantasies and have difficulty distinguishing between fantasy and reality, and tend to keep their fantasy worlds closely guarded secrets."[39]

Fifty-eight per cent of one sample studied by Wilson and Barber reported spending much time in their childhood playing with fantasised people or imaginary playmates. Another 92 per cent saw themselves as "psychic" or sensitive, with telepathic ability and experiences of poltergeists and other phenomena throughout their later lives. A table of 132 alleged UFO abductees drawn up by Bartholemew and Howard found similar correlations with their life experiences.

Many of the percipients, whose claims stretched from the 19th century to the present and involved multiple experiences, reported astral travel, religious and psychic visions, and telepathy alongside their UFO contact. A significant proportion of these described suffering illness and real physiological effects as a result of their 'fantasy' experiences. Others had beliefs in guardian spirits and "spirit guides," whilst a number were assessed as being excellent hypnotic subjects. The researchers concluded that the results of their studies had clear implications for the understanding of many UFO abduction and contact reports, and stated that "the images of the modern-day Fantasy Prone Personality reflect the period and culture into which they were born."[40] Many of the fears expressed in the messages received by the witnesses to aliens and spirit guides concern warnings about nuclear destruction and other apocrypha.

Alex has said: "I enjoy watching the kingfishers diving in the water near me and the abundance of wildlife God has given this great planet for us all to enjoy. I sincerely hope that mankind will reverse the destruction he violates every day to the earth in the name of money, before it is too late for us all."[41]

Conclusions

Like so many others involved in the UFO cottage industry, Alex Birch now has his own website, www.ufo-images.ndirect-co.uk. On this "fantastic site for UFO buffs and everyone else," the web surfer can read about the Birch sighting, see and order copies of the photograph, and purchase copies of the Air Ministry report. Alex's son Adrian advertises quality hand-crafted wooden models of classic UFOs like those reported by 1950s contactees George Adamski and Howard Menger. It is an uncritical site, designed to market the case, but also to inform people about both the original sighting and how Alex sees mankind in the cosmic scheme of things. Echoing the apocalyptic fears expressed by many UFO contactees, he writes: "Perhaps we are in the infancy of our species. We peer into the Dark, fearing it, yet seeking within it a reassurance that we are not alone. Perhaps in the black void are beings not unlike us, but maybe wiser, better, who will tell us secrets that will save Us from Ourselves."[42]

Perhaps ... In 1999 the problems surrounding the Birch photograph are no more resolved than in 1962. Indeed, the case is more complicated, not least because of Alex's claims of a lifetime of paranormal phenomena, experiences shared to some degree by his wife, children and other independent witnesses. If the photo is a fake, then is Alex lying about these experiences, too? If so, are his family also lying? Why would anyone create such a web of deceit around themselves for no discernible reason? Yet what are the alternatives? Questions tumble over themselves in desperation to be asked, but any questions merely beg further questions. Only blind acceptance or outright accusation seem to offer any relief.

Perhaps it is all just as true as Alex and his friends had originally claimed, and five strange light-emitting objects were in the sky over Mosborough that March morning. What then? We are still no nearer to divining what he caught on film. Or maybe the photograph was faked, but was just an outward expression of Alex's saucer-haunted life, in the same way the two cousins faked the famous Cottingley fairy photographs, yet always insisted they did see the real thing. In fact one cousin, many years later and close to death, said she had faked the pictures to prove to others the reality of the beings seen at Cottingley that most people were not able to experience. And the other cousin swore on her death bed that most of the fairy photos were a hoax but one of them was real![43] The supernatural explanations appear to be as incomprehensible as the natural ones.

Ultimately, no one knows the truth behind the photograph besides Alex and his two friends. The original negative is lost and any proof there is now lies in the narratives of the three witnesses. But by their own admission, they have intentionally blurred the line between reality and fantasy, asking, at various times, for both to be accepted as the truth. As investigators in this case, we find ourselves caught up in the dilemma that anything we write will affect what others choose to believe, but won't change what happened on that March day in 1962. So, in lieu of any hard evidence, all we can safely say is that whilst sincerity does not equal proof, nor does suspicion prove disingenuity.

Be warned. If any of your children claim to have photographed UFOs, or any other mythical phenomena, at the bottom of the garden their lives will never be the same again.

But we leave the last word to one of the three witnesses, Stuart Dixon, who said in 1999: "I find it far better and simpler to let people think what they want to about that photo. I don't care anymore."

8

FIRE ON THE MOUNTAIN

The Berwyn Mountain UFO Crash –
A British Roswell?

ANDY ROBERTS

"There was no doubt in my mind that something of great significance had occurred on the Berwyns ... I leaned towards accepting the extraterrestrial hypothesis." – Nick Redfern, *Cosmic Crashes*, Simon & Schuster 1999.

In 1958, author Gavin Gibbons wrote *By Space Ship to the Moon*, a sci-fi book featuring a UFO landing on the Berwyn mountains in Wales. Sixteen years later, in 1974, those same mountains would again be the focus for a story involving a downed UFO. But this time, some said, the account was for real.

The Berwyn Mountains run south-west to north-east across central North Wales, separating Shropshire from the Snowdonia National Park. They have a long history of human habitation. Prehistoric man lived and worshipped on the mountains, leaving behind a dramatic ritual landscape to which many strange beliefs have become attached. UFOs are not new to the area. Local folklore tells us that these peaks have been haunted by a multitude of aerial phenomena, including the spectral Hounds of Hell, whilst to the south, at Llan-rhaedr-y-Mochnant, the villagers were once plagued by 'flying dragons,' a common historical name for UFOs. Contemporary paranormal puzzles abound, too, and besides UFOs include 'phantom bombers,' ghosts and lake monsters. The region is also the lair of that most modern of mysteries, the 'alien big cat.'

Although popular as a tourist destination, the Berwyn Mountains can be highly dangerous and mountain rescue teams are frequently called out to search for the lost and injured. The highest peak, Cader Berwyn, rises to 827 metres. Several aeroplanes, both military and civilian, have crashed on its slopes in poor visibility over the past 50 years. In winter the area is especially remote, often snow-covered, and dark for over twelve hours a day. It is an ideal spot, if ever there was one, for a UFO landing.

It is against this backdrop that an incident took place on January 23, 1974, which perplexed locals and later the UFO research community. The events spawned a cascade of rumours which led some investigators to conclude that an extraterrestrial craft crashed on Cader Berwyn. These same UFOlogists also claimed that the alien crew, some still alive, were immediately whisked off to a secret military installation in the south of England for study and that the whole fantastic business was hushed up by the UK government. The Berwyn Mountain Incident has been described as "… the best example of a UFO retrieval in Britain," and likened to the Roswell and Rendlesham events.

A preposterous claim? Certainly – and a claim easily dismissed by those with little or no knowledge of the case. But there is no smoke without fire. Even the most bizarre story must have its genesis in truth, no matter how mundane or exotic that truth may be.

Imagine for a moment the consequences if aliens really *had* fallen to earth that night in January 1974. If this speculation could be proved, we would know with certainty we were not alone in the universe. The possibilities and consequences of such an event are awesome. Such proof would also demonstrate that the government had been keeping The Greatest Story Never Told hidden from us. Proof of a genuine UFO crash on Cader Berwyn would blow the lid on the alleged world-wide UFO cover-up.

But if it can be argued that there was no alien craft, what lies behind the longevity and tenacity of these claims? Could it have been the crash of a secret military test craft such as one of the 'flying triangles' which have dominated UFO-lore throughout the 1990s? Or perhaps a failed missile test from the rocketry range at nearby Aberporth? A hoax even? Or something far more complicated? And if it *is* any of these, why have the claims of UFOs, alien cadavers and military cover-ups persisted for over twenty-five years?

Comparisons with Roswell and other UFO crash-retrieval events show the Berwyn Incident to have many of the same components and motifs and therefore to be worthy of in-depth study. Yet whilst rumours of this crash have been in existence for a quarter of a century, it has only recently drawn any serious attention from the UFO community. Although dramatic claims have been made, no one had investigated this potentially remarkable case in any great depth. The Berwyn Incident, far from proven, was a kaleidoscope of rumour and fact concerning crashed UFOs, alien bodies, military retrieval teams, earth tremors, meteorites, weapons testing, disinformation agents, Men in Black and geologically created lights.

The Berwyn Incident

The story is a complex one. I have pieced together a composite account from statements and articles by witnesses, informants, UFOlogists and newspapers of what allegedly happened on and around January 23, 1974. This is 'the story,' the generally accepted account, variations on which have become enshrined in UFO literature and seeped out into the public's consciousness.

Before the Berwyn Incident, the north of England had been plagued by an aerial phenomenon dubbed the 'phantom helicopter.' Over a hundred good sightings were made of this anomalous object, which was seen flying low at night, often over dangerous terrain and in appalling weather. These sightings largely took place between spring 1973 and spring 1974 and ceased, coincidentally or curiously, immediately after the Berwyn Incident. Despite the numerous sightings and keen police interest, which led to a still-secret official report, no one explanation was ever found. But *something* was flying around the northern skies. Many of the witnesses concurred that whatever it was, "it seemed to be looking for something."[1]

Wednesday, January 23, 1974, was just another day in Bala and the nearby villages of Corwen, Llandrillo and Llanderfel. UFOs were the last thing on the villagers' minds as Britain huddled in the depths of winter and the recently introduced three-day week. But as night closed in an event took place which was to change all that.

Just after 8.30 pm, thousands of people in the area were jolted from their winter musings by at least one, possibly two, explosions, followed immediately by a terrible rumbling. The whole event lasted four or five seconds. Furniture moved, ornaments rattled, buildings shook. Livestock and domestic animals voiced their terror. As people shot to their windows, some saw lights streaking across the sky. Villagers flooded out into the streets in an attempt to discover the cause of the violent disturbance. As they looked up into the mountains, several saw a mysterious white glow, lasting a few seconds. Others witnessed beams of light being projected into the night sky.

Believing that a disaster of some kind had taken place, many villagers immediately called the emergency services. After speaking to the police, one local nurse was certain that an aircraft had crashed and set off for the mountains in her car, dreading what she might find there, but eager to offer help until the emergency services arrived. Once above the tree line and on the high mountain road, she stopped her car, baffled and startled at what she could see. For there, high on the desolate mountainside, was what appeared to be a large glowing

146

sphere. Whatever it was lay too far from the road to be reached on foot, so all the nurse could do was watch.

The sphere seemed to pulsate, changing colour as it did so from red to yellow to white, while other white lights, "fairy lights" as the witness described them, could be seen above and below it on the hillside. Realising she could not possibly reach the lights, the nurse drove back to her village. As she did so, a group of police and soldiers stopped her, and forcefully ordered her off the mountain, saying the road was being cordoned off.[2]

Official reaction was quick to the initial explosion. Suspiciously quick some say, with more police and military arriving within minutes, turning people away from the mountain roads. In the days following, it seems there was an unusual and large military presence in the area. Roads remained closed and farmers reported they were forbidden from tending their stock. *Something* was obviously being sought or why would military jets and helicopters be crisscrossing the area and strangers combing the mountainsides? Scientists from university departments also came to tramp the hills. But far more suspicious were the official-looking outsiders who turned up in the villages immediately after the event, tight-lipped about their business, but keenly interested in the events on the mountain.[3]

Media response

The incident was immediately taken seriously by the media, with national TV and radio reports being broadcast over several days. *The Daily Telegraph, The Guardian, The Times* and other national newspapers gave the event in-depth coverage as did the Welsh regional and local Press. Speculation about the cause of the explosion, rumbling and lights was rife.[4] An aircraft crash would have accounted for the noise, lights and keen official involvement. Indeed, one local newspaper was certain that whatever had taken place involved a crash of some kind and that something had been retrieved from the mountains, noting, "There is a report that an Army vehicle was seen coming down the mountain near Bala Lake with a large square box on the back of it and accompanied by outriders."[5]

But the authorities steadfastly refused to acknowledge that anything unusual had taken place. And in any case, not one of the 'explanations' took into account the *totality* of what had been reported by witnesses. Meteorites and earth tremors were also suggested as being the cause, and indeed would have explained *some* of the mystery. But what could possibly explain the 'glows' and 'beams of light' seen on the mountain? They were swiftly dismissed as the villagers' imaginations, shooting stars, or, more ludicrously, people out poaching hares.[6] Natural phenomena were also unlikely to lead to roads being closed by the army or large areas of mountainside being prohibited to the public.

With no further information coming to light, the media soon forgot about the incident. The locals, too, let the matter fade from their immediate concern if not entirely from their memories. UFO researchers realised that *something* had taken place which had not been satisfactorily explained. Lights in the sky and mysterious explosions, together with unusual military activity, are avidly noted by the UFO community. However, in 1974 UFO crash retrievals were barely mentioned in UFO literature, especially in the UK, and there was no immediate template for the events in the Berwyns to fit into. Various UFO journals reported the events at the time, but no investigation was undertaken and no real conclusions were offered.

But shadowy forces appeared to be at work. Within months of the event, UFO investigators in the north of England began to receive official-looking documents from a group called Aerial Phenomena Enquiry Network (APEN). These documents claimed that an extraterrestrial craft *had* come down on the Berwyns and was retrieved for study by an APEN crash retrieval team which had been on the scene within hours of the event. More significantly, APEN claimed there had been a key witness to the UFO crash who they were recommending for hypnotic regression. At that time, this was virtually unknown in the UK UFO community. In fact, besides having being used in the 1961 Betty and Barney Hill 'abduction,' hypnosis was not then widely employed within UFOlogy.[7]

If APEN *were* hoaxers, they displayed an uncanny and detailed knowledge of both UFOlogy in general and the Berwyn Mountain Incident in particular. Some researchers have speculated that APEN may have been part of a government cover-up, using UFO mythology to spread disinformation and so divert attention from secret weapons testing. APEN also issued similar enigmatic communications in conjunction with other UFO events, notably the Rendlesham Forest case.

Awakened interest

The Berwyn Incident lay largely dormant throughout most of the 1970s and 1980s, being little more than a footnote in UFO literature. But intriguing pieces of information did surface, later becoming part of the lore surrounding the case. Jenny Randles was by chance a frequent visitor to the region in the late 1970s, staying in the Llandrillo area for weeks at a time. She recalls locals speaking to her about military activity on the mountains in the wake of some form of crash-like event. Jenny became mildly interested in the case, and initially put it down to a possible "earthlight", for that reason posting reports not to UFOlogy but to Paul Devereux, an earth mystery researcher. Indeed Jenny saw a mist-like earthlight rise from rocks in the Berwyns at Llandrillo in 1978.

In *Places of Power*, Paul Devereux briefly relates the Berwyn Incident, attributing the cause of the odd lights seen on and above the mountain to geophysical stresses. Known as "earthlights" to UFOlogists, these are literally lights formed by Earth. Devereux notes that a colleague, Keith Critchlow, was in the area several days after the incident and "fell in with scientists who were investigating the mountain." They had a Geiger counter with them, which allegedly gave extraordinary readings in the vicinity of a Bronze Age archaeo-logical site known as Moel ty Uchaf, on the slopes of Cader Berwyn.[8]

The 1990s brought growing interest in the UFO subject and the Berwyn Incident was revived. Jenny Randles lectured on the case at the 1994 *Fortean Times* UnConvention and mentioned the anomalous radiation count at the Moel ty Uchaf circle. Following her lecture, she was approached by a science corre-spondent from the *Sunday Express*. He mentioned rumours received by the paper of a leukaemia cluster among children in the Bala area which had arisen in the years following the Berwyn Incident. At the time, he connected it with possible leaks from the Trawsfynedd nuclear power station, but there was no evidence for this. In the light of later claims of UFO crashes or secret military hardware, it could be implied that whatever did crash had possibly been radioactive in nature and of sufficient strength to affect the human organism.[9]

By 1996, the Berwyn Incident had featured in UFO books, several UFO magazines and national newspapers. Television programmes on Channel 4 and the Discovery Channel covered the case, and by 1997 it was the focus of an entire chapter in Nick Redfern's best-selling book about the government cover-up of UFO information, *A Covert Agenda*.[10]

The Berwyn Incident was big news once again. From its humble begin-nings, it was now a 'British Roswell' just waiting to burst, firmly enshrined in UFO-lore as one of the United Kingdom's few UFO crash retrieval cases. This surge of publicity brought forward new witnesses, whose testimony added fresh and dramatic dimensions to the case.

Bodies of evidence

In an article for *UFO Magazine*, veteran UFOlogist Tony Dodd recounted how his anonymous informant was part of a military unit put on stand-by several days *before* the date of the Berwyn Incident. His unit was moved northwards through North Wales until he and four others were sent to the village of Llanderfel to collect "two large, oblong boxes." They were ordered to take these to Porton Down in Wiltshire.

Once at Porton Down, a UK government research establishment, the boxes were opened. Dodd's informant told him: "We were shocked to see two creatures which had been placed inside contamination suits. When the suits were fully opened, it was obvious the creatures were clearly not of this world and

---

CIRCULAR LOGIC

In August 1980, Wiltshire farmer John Scull found rough circular swathes of flattened oats. Mildly puzzled, he reported the matter to local Press. Speculation that a UFO had created 'saucer nests' was soon rife. But Bristol-based UFO group, Probe, and a Bradford-on-Avon-based physicist, Dr Terence Meaden, suggested the most likely cause was the weather.

Every summer brought a handful of new circles, mostly to the counties of Wessex. In the 1960s, strange lights had been seen nearby at Warminster, making it a Mecca for UFO enthusiasts.

In July 1983, the phenomenon escalated when the national Press discovered its photogenic qualities. Immediately, hoaxers were caught in action. One national newspaper paid a farmer to fake a five-circle pattern in the unfulfilled hope of catching out a rival.

BUFORA ran a battle to persuade the media and

public that circles were both weather effects and trickery. In 1986, they staged a London seminar to make this point, a massive vote of UFOlogists in the audience endorsing the solution.

Hampshire-based statistician Paul Fuller, and Jenny Randles, Director of Investigations, published the first-ever crop circle report, *Mystery of the Circles*. Although this had some effect, circles continued to grow in numbers and complexity as if someone was playing to the crowd.

From simple circles in 1980 to fantastic geometric designs that resembled whales and spiders by 1990, patterns in the crops grabbed world attention. Publication in 1989 of a glossy book with aerial photos by Pat Delgado and Colin Andrews began a circus. New Age travellers, mystics, UFOlogists and sightseers arrived in their thousands each summer to hunt for patterns. The belief was that these marks were a cosmic message warning the earth of ecological disaster. This caught the

cultural ethos.

Circles by now had spread all over the world, but more than a hundred a year still made Wessex the hub. Frantic efforts by BUFORA and meteorological research team TORRO to bring sanity to the debate were lost as circle research groups, magazines and conventions sprang up. Many farmers got angry at the trespassing and damage to their fields. Others judiciously faked circles and charged entrance fees, realising more could be made this way than by selling their wheat!

In September 1991, two elderly Hampshire artists confessed they had hoaxed hundreds of circles since 1980, coming up with ever more complex shapes to keep the Press and public happy. Media lost interest for years, but circle research went on regardless and remains very active. Indeed in 1999 new books and media interest began.

Circles still appear in numbers. The millionaire Rockefeller Foundation when examined were found to be dead. What I saw in the boxes that day changed my whole concept of life." Dodd's informant goes on to relate details of the creatures – "The bodies were about five to six feet tall, humanoid in shape, but so thin they looked almost skeletal with covered skin."

The military man did not actually see a crashed UFO himself, but claimed that "Some time later, we joined up with the other elements of our unit, who informed us that they had also transported bodies of 'alien beings' to Porton Down, but said their cargo was still alive."[11]

announced a large grant in May 1999 to Colin Andrews to research the mystery on their behalf.

Is it real?
● There can be no question that circles do exist. Thousands of photographs of simple circles, rings and complex shapes are on file. Over 30 countries have recorded them, but 90 per cent formed in Britain.
● Circles are best observed in cereals, but have formed in grass, snow, sand and wet road surfaces.
● Simple circles have been traced via scientific records, old woodcuts and aerial photographs back to at least the 16th century. No known complex shapes were found before the 1980s.
● Several good eyewitness accounts describe the formation of circles (simple ones only).

Is it solved?
● Nobody has ever reliably reported the formation of complex patterns or alien UFOs creating crop circles (merely lights). Film evidence of such things since 1996 has crumbled on investigation.
● It has been demonstrated many times that circles can be hoaxed well enough to fool even experts. Methods used include ropes and flat boards, acting like snowshoes to spread the pressure.
● Weather archives show that electric whirlwinds can create patterns in fields and suck up debris as a disc-like cloud over the circle. The hoaxing artists admit they got the idea from real circles produced this way in January 1966 amidst snake-infested reed beds south of Cairns, Australia.
● Experiments by physicists in Japanese universities have recreated mini-whirlwinds and produced artificial circles in the lab. They then sought locations where rotating winds may create circles in dust and found dozens in closed tunnels of underground railway systems in several major cities.

Conclusion
Despite all the media hype and the waste of money by bodies such as Rockefeller, crop circles are not a mystery. They were resolved years ago. The vast majority – certainly the complex shapes – are hoaxes created by many people who have fun with researchers and the media each summer. It has become a game.

Behind these are a few, simple circles that have appeared world-wide for centuries and result from electrified whirlwinds. The Japanese have quickly realised these can be harnessed for energy purposes. The rest of the world foolishly chases alien messages in crop fields. "You are mugs," would be quite an apt one.

Further reading
Crop Circles: A Mystery Solved. Fuller, P. and Randles, J. Robert Hale, London, 1993.

Paper chase

This interest by the media, together with claims made by researchers such as Jenny Randles, Nick Redfern, Tony Dodd and Margaret Fry, led to me re-investigating the Berwyn Incident in 1998. As there was a wealth of information available, I reasoned that somewhere, amid the accounts of the witnesses and the claims of UFOlogists, lay the key to what really happened on that January night in 1974.

UFOlogists, particularly those who believe that there is a global conspiracy to conceal evidence of extraterrestrial visitation, are keen to stress the importance of the "paper trail." By this, they mean that any event, however secret, must have generated some official documentation, and that by finding it, clues as to what happened can be gleaned. It seemed reasonable that an event of the magnitude of the Berwyn Incident would have left at least *some* trace in official records, no matter how small or obscure. But those UFOlogists who had pursued the case up to 1997 had not followed this line of enquiry, some claiming that either the documentation no longer existed or was part of the cover-up. They clearly had not looked hard enough because I found a wealth of official documentation from a variety of sources. I used it, together with witness statements, to piece together the true events of January 23, 1974. What follows is the result of that re-investigation.

In *A Covert Agenda,* Nick Redfern suggested that the numerous "phantom helicopters" seen in the months leading up to the Berwyn Incident were flown by military UFO crash retrieval teams. Redfern also claimed they had received advance knowledge of a UFO landing and were on permanent stand-by, suggesting that "Perhaps the idea of a joint CIA–Ministry of Defence project designed to respond on a quick reaction basis to UFO incidents should be considered."[12]

But the phantom helicopter story is a red-herring. Although a number of people had called the phenomenon a "helicopter," a motif quickly seized upon by the media, most witnesses were in fact describing an unknown *light* of many shapes and colours. The "phantom helicopter" was more Unidentified Aerial Phenomenon than Unidentified Flying Object, a big difference. Some genuine helicopters were proved to be responsible for certain sightings, but the rest remained unexplained.[13] Additionally, the phenomenon was not seen in the Bala area. There is no real connection between the "phantom helicopters" and the Berwyn Incident other than the circumstantial link made by Nick Redfern.

During my research into the Berwyn Incident, I discussed this in some depth with Nick Redfern, who still stands by his published link between the "phantom helicopter" and the Berwyn Incident. But in correspondence, he qualified his belief with "All I was really trying to do was get people thinking about what *might* have taken place – nothing more."[14]

A strange night

January 23, 1974, was a strange night by anyone's standards. In retrospect, it was one of those evenings when nature was staging a *son et lumière* display on a scale rarely seen. Witnesses in the villages surrounding the Berwyn Mountains reported seeing a great deal of aerial phenomena that night. Besides the odd lights seen on the mountain, their reports and those of the media describe at least four incandescent balls of light which streaked across the Welsh skies between 7.30 and 10.00 pm that night. These sightings have been seized upon by UFOlogists, the implication being that what was seen were UFOs, at least one of which crashed or landed on Cader Berwyn. To the villagers of north Wales, they *were* UFOs – literally Unidentified Flying Objects – and they described them in terms which make them sound highly unusual.

"I saw this object coming along the mountain, about the size of a bus really, white in the middle," said one farmer. "It came across the mountain and dipped. I thought it was going to crash."

This dramatic description certainly sounds like many UFO accounts. But there is a rational explanation for Farmer Williams' sighting and all the other aerial phenomena seen that evening.

Records kept by the Astronomy Department at Leicester University, among other places, show that a number of outstanding bolide meteors were seen that night. These coincided with the approximate times given by witnesses in north Wales. The first was at 7.25 pm, followed by another at 8.15 pm. The third, at 8.30 pm, coincided with the centre-piece of the evening's events. And yet another, the most dramatic of all, was seen at 9.55 pm.[15]

Bolide meteors are considerably brighter and longer lived than ordinary 'shooting stars.' They can appear to be very low, depending on the position of the witness, and often trail 'sparks' of blue and green across the sky. Such meteors are responsible for many misperceptions of UFOs and even fool the emergency services, who are often called out to 'plane crashes,' only to discover the witnesses had seen a bright bolide meteor.

At exactly 8.38 pm, the Bala area was rocked by a huge explosion, closely followed by a deep rumbling. One witness recalled it as being "like a lorry running into a house." Crockery rattled, furniture moved and walls rippled slightly. Some people were certain it was a plane crash on the mountains. Other, older residents of the area, recalled earth tremors of the past and assumed it was the latest in a series of such disturbances which have taken place along the geological rift known as the Bala Fault.

This is the primary incident which subsequently caused many UFO investigators, and the readers of their books and articles, to suggest and believe that a UFO crashed. In effect, they are saying that the noise heard and impact felt was the UFO impacting on Cader Berwyn. However, the crashed UFO story only came out years *after* the event. At the time, confusion reigned as to what had caused the impact.

Because of reports of lights in the sky that evening, it was initially thought that a meteorite had impacted on the Berwyns. Many people across North Wales claimed to have seen a light in the sky "trailing sparks." But this was at 8.30 pm, eight minutes before the explosion, and witness descriptions indicate that it was yet another bright fireball meteor. Nonetheless, in the minds of many it has become conflated with the 'explosion' to create evidence of a crash.

The explosion was heard only in the Bala area, but the tremor was felt as far away as Liverpool. By 2 pm on January 24, seismologists had determined that the explosion and tremor were caused by an earthquake of 4–5 on the Richter scale.[16] Its epicentre was the Bala area at a depth of eight kilometres. To cause a reading of that magnitude, a solid object – meteorite or UFO – would have weighed several hundred tons and left a massive crater. Therefore, unless a UFO had crashed at the *exact* moment of an earth tremor, it can be safely assumed that the explosion and rumblings were the result of a purely natural process.

A nurse's story

Following the explosion, Llandrillo district nurse Pat Evans ran out into the village street. She saw no lights, but the explosion and the accounts of other villagers convinced her that something had crashed on the mountains. It took her a while to get through to the police as the phone lines were jammed with 999 calls, but eventually she spoke to Colwyn Bay police HQ. They suggested it could have been a plane crash, so she bundled her two young daughters into the car and set off up the mountain, intending to offer help until the emergency services arrived.

As Mrs Evans reached the point where the B4391 mountain road levels out, she was puzzled by what appeared to be a large illuminated ball of light on the hillside. Unable to identify it, she drove on for a few minutes before returning to the same spot. The light was still there, so she parked and observed it for a while. A light drizzle was falling, but the night was otherwise clear. Mrs Evans described the ball as "large," and forming a "perfect circle," but it did not appear to be three-dimensional.

In an interview she recalled: "There were no flames shooting or anything like that. It was very uniform, round in shape ... it was a flat round." As she watched in puzzlement, the light changed colour several times from red to yellow to white. Smaller lights, "fairy lights" in Mrs Evans' words, could be seen nearby. It was too far away to reach on foot, so she returned home to bed.[17]

Writing about the Berwyn Incident, many UFOlogists have claimed that Mrs Evans was turned back from the mountain by soldiers and police. This is untrue, and arose from a misunderstanding when she was first interviewed by UFOlogists. Pat Evans is furious she has been misrepresented in this way, and stated unequivocally to me in 1998 that she saw "not a living soul" on the mountain that night. More importantly, a letter from her exists, pre-dating any interview, noting that she saw no one. This fact is significant because the misreporting of Mrs Evans' experience has lent credence to claims that a crash retrieval team was on the mountain shortly after the explosion.[18]

The BGS study

Nonetheless, what the nurse saw on the slopes of Cader Berwyn was still crucial to any explanation of the case and I wanted further evidence untainted by time or UFOlogists. For that evidence, I turned to records kept by the British Geological Survey in Edinburgh. The BGS records, untouched for twenty-four years, revealed that within days of the explosion a team of investigators had been sent to the Bala area. This, incidentally, is almost certainly the source of rumours of 'officials' who came to the area, stayed in local hotels and questioned villagers closely about the event. That is exactly what the BGS field team did.

A total of six interviewers went to the area and conducted door-to-door enquiries about the event. This is the procedure by which the BGS investigates earth tremors and earthquakes. The interviewers worked to a set questionnaire, which asked questions such as, "Were you at all alarmed or frightened?" and "Did you hear any creaking noises?" These and similar questions must have seemed quite odd to the locals, especially when asked by a team of outsiders who just arrived from nowhere. Over two hundred witnesses were interviewed. Nurse Pat Evans was one of them.

The BGS field notes were enlightening. Most UFOlogists have always assumed that Pat Evans must have been on the mountain almost immediately after the explosion. They use this assumption to argue that the lights she saw surrounding the anomalous red lights she reported must have been from a pre-alerted crash retrieval team as no one else could have got on the mountain so quickly after the 'crash.'

But the BGS records from her 1974 interview are very specific about time and say she "left the house during 'Till Death …'." I took "Till Death" to be a reference to the popular TV sit-com *Till Death Us Do Part* and checked TV schedules. Sure enough, *Till Death Us Do Part* started at 9.30 that night. *Till Death* … was the only post-8.30 pm sit-com that evening. Knowing that Pat Evans left the house after 9.30 pm means she would have observed the anomalous light some time after 9.40 pm, an hour later than some had previously thought. But that hour's difference is crucial.[19]

Mountain rescue

Meanwhile, 14-year-old farmer's son Huw Thomas was also watching TV that night. At about 9.20 pm, he answered the door to find several policemen in the farmyard. They wanted to commandeer the farm Landrover, saying a plane had crashed up on the mountain. Thomas' parents were out so, with his neighbour Enoch driving, they set off up a track leading to the mountain, other police following in a car. Nearing the mountain-gate, they had to waste valuable time moving a car which blocked the road. Huw Thomas recognised the car as belonging to local poachers. Once through the mountain gate, several policemen spread out on foot with torches whilst the Landrover and police car drove slowly up the track.[20]

The time it took Huw Thomas to speak to the police, load the Landrover, drive up to the mountain and move a car from the road would place the police search team on the lower slopes of Cader Berwyn at about 9.40 pm.

The BGS also interviewed one of the poachers whose car Huw Thomas had moved. This interview confirmed their time and position, and states that the poachers "carried on work for forty-five minutes (after the explosion) and were almost back at the car when they met party (police, etc) coming up."[21] Huw Thomas, now a farmer in his own right, confirmed this meeting in a 1998 interview.[22]

That the search party comprising police and farmers met the poachers as they went up the mountain is further backed up by other BGS material. Besides interviews, the BGS records also contained an Ordnance Survey map on which important witness locations and sightings of lights were plotted. This map was a revelation. It showed the anomalous light seen by the nurse, the location of the poachers and the police search party to be *all in the same small area of hillside*. And, as already noted, the times given to the BGS by all three parties place them there *at the same time*.

The logic and conclusion are inescapable. Neither Huw Thomas nor the police saw the light seen by the nurse. Conversely, the nurse *did* see the police, though she did not realise it at the time. The drawing on her BGS notes clearly shows and describes "vehicles" and "torch lights." This was the search party. Between them, very close to both, is the anomalous light source. Whatever she was seeing *must* have been visible to the search team and the poachers. So either the farmer and police lied about what they saw to the BGS in 1974 and myself in 1998, or it was not noteworthy at the time.

Illumination

But what was it? Well, there is one possibility which would account for it. The BGS notes also confirmed that the poachers were using powerful lamps made from car spotlamps powered by car batteries. Pat Evans recalls the weather was clear but drizzling. Lights seen in those conditions can appear to change colour and size by refraction and to 'glow.' As for the size, which she described as larger than vehicle lights, this may be a perceptual trick. Remember that Nurse Evans was looking across a dark mountainside with no visual points of reference and expecting to see a plane crash or some other scene of devastation. On the evidence available, it is certain that the nurse saw the poachers with their lamping lights at the point they met and talked to the police.

Some UFOlogists claim that although bolide meteors *were* seen throughout the evening, the beams of light on the mountain immediately after the explosion were not astronomical in origin and were connected to the UFO crash. Several of the BGS notes refer to people seeing these beams "on the brow" of the hill, "sometimes on and sometimes off, but always vertically into sky." Another witness saw one beam "processing about the vertical." These accounts were puzzling until I looked closely at the locations of the witnesses.

All the witnesses who reported seeing these 'light beams' were in the village of Llandrillo at the time. The land rises sharply to the south. To an observer in the village, the 'brow of the hill' is not the summit ridge of the Berwyns (actually over three miles away), but the plateau area around the 548m point – the exact area, in fact, where the poachers with lamps were. The BGS records note that the poachers "continued work for half an hour to forty-five minutes" after the 8.38 pm earth tremor, and it was early in this time period the beams were seen. Some villagers were convinced that poachers' lamps could not be responsible for the beams, others not so sure. One witness told the BGS he had seen the poachers' lights on previous occasions and they were exactly the same as the beams that night.

This theory may appear to be debunking or twisting the facts to fit a theory. But we must use logic and probability in solving any case. The facts are that poachers with powerful lamps were in the *exact* area where the beams of light were seen. When questioned by police, the poachers claimed their lamps were not responsible, and that they had kept them trained on the ground. Yet they also said they had not seen anything unusual. It is reasonable to suggest that as the poachers and their bright lamps were in the same location as the beams of light seen from Llandrillo, it was their lights people saw and misperceived, perhaps because of excitement caused by the earth tremor, perhaps because of belief in a crash of some kind.

The poachers had very good reason for not wishing to own up to causing bright beams of light in the sky as it was reports of 'light beams' which partially led the police to believe an aircraft had crashed. However, there *were* a very small number of genuinely unexplained lights seen that evening. One witness opened her curtains immediately after the tremor to see a "big bright glow in the sky over the brow of the hill." To the south-east, another saw a "glow several times brighter than the sun," which "came and went." Maria Williams of Llandrillo noticed this white glow at the same time as the poachers' lights. Other witnesses reported a fire-like glow, lasting a few seconds.

Some scientists have suggested this short-lived white glow was caused as a result of the huge tectonic stresses involved in the earth tremor – an earthlight. But witnesses to this were few – and as it was seen at the same time as a bright meteor and the poachers' lights, it may well be yet another misperception. Indeed, one witness described the 'glow' as "twinkling ... like a streetlamp seen through heavy rain," just how a bright lamp would appear.

Military presence

Claims by UFOlogists that a military presence was on the scene immediately following the 8.38 pm explosion and in subsequent days also bear close examination. As we have already seen, nurse Pat Evans, by her own admission, was not stopped by soldiers or police and saw no one out on the mountain roads. She set off at 7.00 am for work the following day and saw nothing unusual in the village. So how did stories of a massive police and military presence arise? To understand that we need to return again to the official records.

Following the 8.38 pm earth tremor, the police opened a Major Incident Log. This shows that police initially thought a plane had crashed, so fire and ambulance services were put on stand-by. At 9.09 pm, the police contacted RAF Valley Mountain Rescue Team (VMRT), based at Valley on Anglesey, some seventy-five miles away.[23] A three-man team left Valley at 9.20 pm, arriving at Llandrillo at 00.10 am. The VMRT log lists the incident as "Unidentified lights and noise on hillside," and comments: "VMRT requested to investigate lights and noise on hillside. Advance party covered relevant area with negative results. Incident produced much local excitement." The fact that VMRT only deemed it necessary to send a three-man team argues strongly against the event being of any significance. On their arrival in Llandrillo, the mountain rescue team consulted with local police, who suggested they wait until morning before initiating a search.[24]

At 7.00 am on January 24, VMRT, together with local police, searched the mountains. They found nothing and abandoned the search at 2.15 pm, possibly following official notification that the 'explosion' had been caused by an earth tremor. Neither the police nor VMRT logs mention any military involvement other than the RAF Mountain Rescue Team. Farmer's son Huw Thomas was again out on the Berwyns that day, acting as guide for Ron Madison, a scientist working on the theory that a meteorite may have impacted. Madison and Thomas recall seeing no one else on the mountain other than the police and VMRT. However, the intense media interest led to various helicopters flying over the area throughout the week, and Ron Madison used his contacts at RAF Valley to overfly the area in a plane to take a series of photographs.[25]

But this low level of official activity would not account for reports of closed and guarded roads, the military presence, or for the aircraft and twin-engined helicopters seen overhead. Looking at the paper trail, none of the police, Mountain Rescue Team or British Geological Survey documents from 1974 mention this alleged military activity. In fact, the only contemporary record of a military presence comes from the article in the *Border Counties Advertiser*, which is the source of rumours of bodies being brought off the mountain.

In looking for an explanation to this component of the story there are two crucial factors. Firstly, none of the Berwyn Mountain Incident witnesses was formally interviewed by UFOlogists until at least twenty years after the event. Secondly, there had been at least one other event in the locality which contained all those elements.

On February 12, 1982, an RAF Harrier jet carrying top-secret equipment crashed on Cader Berwyn. The RAF descended on the area in force, using Gazelle and Wessex helicopters, together with Harrier and Hercules planes, in the search. The tiny village of Llandrillo – the centre for this activity – was alive with RAF trucks and personnel for several days. The crash site was sealed off and guarded until the wreckage could be removed.[26] Additionally, there was another crash of a military plane, also carrying top-secret equipment, on the same mountain in 1971 or 1972, two years before the alleged UFO crash. Again, the area was sealed off with a large military presence. It is almost certain that these incidents, at the same time of year on the same mountain, were confused with the 1974 events.

Crash retrieval

"But," believers in a genuine alien crash say, "what about the military inform-ants who came out of the woodwork in 1996 claiming intimate knowledge of, and participation in, the crash retrieval?" Initially, this strand of the story seemed promising. After all, when ex-military men are speaking out surely there must be *something* in their story?

However, these 'military informants' who contacted researchers Nick Redfern, Margaret Fry and Tony Dodd did so only *after* the story had been in a 1996 issue of *UFO Magazine*. They fuelled the controversy surrounding the story, offering much speculation, but no verifiable fact. Redfern recently told me that his informant's telephone number is 'dead' whilst Dodd refuses to expand on the identity or veracity of his contact. A close reading of Dodd's account throws up more questions than answers. If the military had obtained aliens, alive or dead, would they really ferry them by truck? Surely a helicopter would have been the fastest, most efficient and secret form of transport. Porton Down, the research establishment to which they were taken, would hardly compromise security or contamination by opening the boxes in the presence of what were essentially delivery boys. Until these UFOlogists can back their claims up with some substantial proof they remain unsubstantiated anecdotes, interesting but inconsequential to the solution of the case.

These 'revelations' came also at a time when several UK UFOlogists were being contacted by alleged 'military sources' offering secret UFO-related information, none of which amounted to anything tangible. Researcher Kevin McClure suggested that this was a well-organised hoax, basing his suppositions on the number of contacts made within a short time-span and the absolute absence of hard proof. APEN, the organisation which circulated pseudo-official documents following the Berwyn Incident, are widely regarded by most serious UFOlogists to have been a hoax perpetrated by UFOlogists on UFOlogists. This sort of hoax is not new to the UFO community, the most famous of the hoaxed documents being the MJ-12 papers, which fooled UFOlogists for over a decade.

Despite the wealth of evidence to the contrary, Jenny Randles is not convinced that the Berwyn Incident is completely solved. She cites the alleged anomalous radiation readings, the conviction of locals in 1978 who told her of a military follow-up and the rumour of a leukaemia cluster reported by the Bala doctor as possible evidence that the incident may have been a military accident, perhaps involving a mishap with a nuclear missile. Such events have been alleged elsewhere. Yet there are problems with Jenny's interpretation. The radiation readings taken at the Moel ty Uchaf circle in 1974 were a one-off. To have any scientific relevance at all, a series of Geiger counter readings prior, and subsequent, to the 1974 event would be required. As for the alleged leukaemia cluster, there is no evidence to support this. Enquiries at the records of the National Radiological Protection Board, Greenpeace, a former radiation monitor at the Trawsfynnyd Nuclear Power Station and the archives of local papers did not reveal so much as a hint of such a cluster which the doctor claimed he found retrospectively from the number of childhood cases he came across. He was completely unaware of the UFO story when making his suspicions public.

Conclusions

In his latest book, *Cosmic Crashes*, UFO researcher Nick Redfern again writes about the Berwyn Mountain incident. Despite the hype surrounding his book, Redfern can offer no new evidence in support of his quote at the head of this chapter. He further muddies the facts surrounding the Berwyn case by inferring a connection with another alleged UFO crash incident which took place in

Staffordshire, on Cannock Chase, a few days prior to the Berwyn event. This tenuous link clearly demonstrates how a UFO case, particularly an alleged crash retrieval, will mutate and grow like a virus, constantly accreting new information and theories, if not actually any hard fact. There is no better subject for the study of living folklore than the genesis and development of a UFO crash retrieval myth.[27]

So that is where the Berwyn case stands in 1999. There are still a few loose ends and niggling uncertainties, but the symmetry of any UFO case is rarely complete, especially when it is not properly investigated for twenty-five years. However, I think the account I have given is the best – dare I say it? – 'explanation' for the disparate events which coalesced into the Berwyn Mountain UFO Crash. Of course, there are those who will still choose to believe an extra-terrestrial UFO crashed on Cader Berwyn. They will continue to insist that documents have been falsified, that witnesses have been misquoted and that a gigantic cover-up exists. That, of course, is their prerogative. But they are conclusions based on belief and not the results of investigation.

My conclusions are based not on belief, but on the 'paper trail' left by police, RAF, VMRT, BGS, and statements made by witnesses. The pattern which emerged from studying those sources is largely consistent, each source tying-in with the others, and those in turn matching witness testimony. I can only conclude that until some hard, consistent, evidence is produced to the contrary the notion that an alien spacecraft crashed in the Berwyn Mountains is redundant.

It is hard to believe that a concatenation of prolific meteor activity, an earth tremor and poaching activity could lead to the conclusion that a UFO had crashed. It did. But sometimes – often, in fact – the truth about a UFO case is far stranger than any fiction. Although I have been investigating mysteries for twenty years now, every case teaches something new or reinforces some basic principle. The Berwyn Mountain case taught me (again!) never to trust material originated by UFOlogists, but always to go back to source documents and witnesses, and try to reconcile the two. It also taught me (again!) about the flaws of perception and of the care needed in interpreting witness statements. However certain a witness may seem, memory often combines disparate events and speculation into a convincing reality.

Charles Fort, the indefatigable researcher and inspiration behind *Fortean Times* magazine, had much to say about the connections – or non-connections – between earth tremors and meteorites. And it may be that there are other, deeper factors at work in the Berwyn Incident. Perhaps earth tremors and bolide meteors are in some way connected by mechanisms at present outside our understanding. Or perhaps extraterrestrials have learned how to enter Earth's atmosphere under cover of meteor showers, even *disguised* as meteors. The adventurous believer might even wish to accept that aliens may even have prescience of earth tremors and be able to effect a landing at exactly the same time. In lieu of hard facts, the speculative possibilities are as endless as they are futile.

On the other hand, it could all be a gigantic cosmic coincidence, a tangle of belief and wishful thinking from which UFOlogists have spun yet another saga in the continuing extraterrestrial mythos.

THE TRANS-EN-PROVENCE UFO LANDING

Late in the afternoon of January 8, 1981, Renato Niccolai heard a whistling sound and saw a grey, disc-shaped object landing in his garden. From a closer viewpoint, he could see the object was "in the form of two saucers upside down, one against the other" and resting on several bucket-like protuberances. After a few seconds, the whistling sound started again, the UFO took off and flew rapidly to the north-east. Niccolai discovered "skid marks" at the landing site.

At first, his wife thought it was a joke when he told her, but after seeing the landing traces took him seriously and told their neighbours, who alerted the police. They came the next day and took soil samples from the landing site, which consisted of two concentric circles and black marks on the ground. Later, investigators from GEPAN, France's governmental UFO investigation unit, took plant and soil samples.

Is it real?

- GEPAN considered the case to be the most significant it had investigated.
- GEPAN's analysis suggested that pressure and radiation had caused changes in both soil and vegetation.
- Inexplicable ground traces suggested something had been at the landing site.

Is it solved?

- The methodology of plant and ground trace analysis was not consistently carried out.
- The ground traces may have been caused by a cement-mixer and vehicle tracks according to a later study of the case by French UFOlogists.
- Niccolai told one investigator: "There are many silly people in the world. On some future day, I shall tell you the whole truth."

Conclusions

Often touted as one of the best examples of a Close Encounter of the Second Kind (CEII), the Trans-en-Provence case actually rests on the shaky testimony of its sole witness. Statements from Niccolai have seriously undermined his credibility. When his wife returned home that day, he told her: "Your cat is back. Extraterrestrials brought him home." On a TV show about the case, he said: "Maybe I saw something. Maybe it's a story ... I say, I too, during the night I dream."

Further reading

"Trans-en-Provence: When Science And Belief Go Hand In Hand." pp. 151–159. Maillot, E. and Scornaux, J. *UFO 1947–1997.* Ed. Evans, H. and Stacey, D. John Brown Publishing, 1997.

9

RENDLE SHAME FOREST

JENNY RANDLES

"This is now regarded as the best-attested case of a UFO crash outside the US."
– Interview with Nick Pope, ex-MoD UFO desk officer,
Mail on Sunday, July 2, 1995.

You may not be aware of the fact, but a small triangular craft smashed its way through pine trees in East Anglia in December 1980, leaving indentations on the ground and excessive radiation in its wake. The effects were so obvious that the forest was quickly felled on orders from the UK government. The many witnesses include officers from the US Air Force, whilst its passage across Suffolk was tracked by MoD radar. A dog was even killed by its deadly energy trail.

An enormous security furore erupted after this case. During a pursuit across miles of British soil, laser beams from a hovering UFO were fired down upon the bunkers where atomic missiles were covertly stored. The outcry almost ended the centuries-old special relationship between Britain and the USA as this astounding case became a political football.

Or, then again, none of these things 'really' happened. It was – as astronomer and UFO sceptic Ian Ridpath has said of the debacle – a "ghastly embarrassment to British UFOlogy."[1] Who on earth is right?

Remarkable events of some sort occurred in Rendlesham Forest, Suffolk, over a dark winter weekend. They were located right in the middle of one of the most security-conscious areas of Europe. The affair reflects precisely what the UFO phenomenon is about for most people – the belief that alien craft can defeat earth's feeble defences and run rings around our military might and civil liberties.

It is easy to see why if just part of this story is true it should entrance UFOlogists. However, like all big cases, strength of evidence does not equate with omnipotence. Rendlesham, as it is colloquially known, has been careering down a slippery slope since 1997 from a once hallowed status as the top British case. It has now reached the point that even some of its proponents wonder if behind the façade lurks shame, not mystery. If so, will it crash back with vengeance on the UFO community? For UFO enthusiasts admitting to a mistake can seem worse than aliens invading our planet. It is done all too rarely.

Time continues to unravel the threads that weave this story together. En route, the saga of Rendlesham Forest illustrates many difficulties faced by a would-be UFO detective. Help from John Le Carré or Stephen King might also not go amiss in trying to figure out this one.

Night of the UFOs

Christmas 1980 was a real 'night of the UFOs.' From soon after sunset on December 25 to around 3 am into Boxing Day the skies above northern Europe were filled with flaring lights. It was cold in the UK, with temperatures hovering around freezing, and the skies were clear. This allowed many people – especially across south-east England – to see the dazzling spectacular that nature put on show.

The fireballs that crossed the heavens were short-lived but brilliant. They were seen from Kent as trails of white light that broke up into pieces. Air traffic flying in and out of both London Heathrow and Stansted Airport in Essex reported near misses. These were certainly exaggerations. The fireballs – astronomers presumed – were a mixture of natural rocks burning up as they entered our atmosphere and the last stages of a Soviet rocket booster that had launched a Cosmos satellite into orbit. None of these things was a direct threat to the airliners, being miles high in the atmosphere, but they undoubtedly were a scary sight when 'up there' and vulnerable in a Boeing 747.[2]

Easily the most spectacular of the events – at 9.08 pm – was of some intrigue to astronomers. They had accepted this to be the death-throes of the booster from Cosmos 749, which was predicted to burn up at some point on its orbit that night. But when they plotted its entry path into our atmosphere something was amiss. Its flightpath altered – not possible for a piece of space metal that smashes towards extinction at the hands of immense friction. The only way an inert booster rocket could change inclination during its dying moments was if something caused the trajectory to be deflected. More to the point, there were reports of debris – assumed to be from Cosmos – reaching the ground in mainland Europe. Yet the witness stories from Eastern England clearly implied that it had burnt up utterly over the Suffolk coast and fell in the sea off Clacton. How did this all fit together?[3]

One day we will need to deflect a huge asteroid heading our way. If we can kick it off course, it might bounce into the atmosphere and leave us alone, otherwise disaster raining down from the solar system is inevitable, as portrayed in science fiction blockbusters such as *Deep Impact*. Earth's orbit takes us through a veritable scrapyard of dangerous objects, and now and then one of them pierces the atmosphere with devastating consequences. In June 1908, a small comet laid waste to hundreds of miles of forest in Siberia with a blast akin to the Hiroshima atom bomb. Had the timing differed by two hours, it would have wiped out Moscow. A few hundred metres bigger and it would have made most of Europe uninhabitable until the next century. This is not even a unique event in the past 100 years.

Today there are serious plans to prepare for the next inevitable disaster, which could be a century from now or tomorrow night. All we know is that it will happen sooner or later. In 1980, nothing was known that could deflect the orbital re-entry of a small rocket, let alone an asteroid the size of Mount Everest. But if the Christmas lights were just the last rites of Cosmos 749, something caused the impossible to happen at around 9 pm on December 26.

East coast wave
At the time of this astronomical riddle, Britain was living within what UFOlogists call "a wave." Some quite remarkable close encounters had occurred across Eastern Britain during a six-week period of the quality and quantity that might normally be expected only over a several-year spell. Such concentrations are a feature of UFOlogy and there are many theories to account for them. They range from the sociological (one person reports a case and this encourages many witnesses to do the same) to the physical (that natural energies may build up within the atmosphere and vent themselves by creating UFOs). Here are some examples of the encounters in the 1980 "wave", although it took months for them to be brought together.

The activity grew from September 13 when police patrols chased a diamond-shaped UFO at Wetherby, and again near the coast at Hull. There were claims by police that the early warning radar unit at Fylingdales detected an unknown intruder that night. UFOlogists discovered that Venus was then very bright in the eastern sky. Had the patrols been fooled by its glare?

That autumn, strange lights were seen in the small villages and hills around Menwith Hill, a highly secret base operated by the American intelligence unit NSA (National Security Agency). This employs electronic and satellite surveillance. Menwith is one of its biggest sites on foreign soil, brooding eerily over the moors with large golfball radar domes. On November 23, a very bright star like a meteor fell from the sky over this area, then headed towards Otley. Witness Nigel Mortimer – who was so intrigued by this case that he became a UFO investigator – was one of several to see it. He says that at first it was assumed to be a 'fireball meteor,' but it suddenly decelerated, changed course and hovered as a blue oval mass. It then surrounded itself in a sort of mist or cloud and headed off in the direction of Fylingdales.

On November 25, a former Royal Navy Commander, then manager of the Chevron Oil Rig in the North Sea, reported how "a glowing orange ball" appeared above the drilling platform as if surveying the site. The next closest rig was twelve miles away, but staff there could also see it. An RAF Nimrod search-and-rescue aircraft was sent aloft as the UFO was now on radar. When it reached the rig, the UFO had gone.

Four days later, Todmorden police officer Alan Godfrey had a close confrontation with a swirling, rotating mass as he drove in his patrol car down Burnley Road. The UFO he saw appears to be real. It caused the trees on the roadside to shake in its wind vortex and even spin-dried the wet road surface directly underneath. Several other police officers chasing stolen motorcycles on nearby hills were amongst locals who also saw the UFO. These are demonstrable, objective facts, although the Godfrey case is best-known because the constable later underwent regression hypnosis and recalled a dream, fantasy or memory (he was unsure which). In this, he was probed inside a weird room by a strange bearded alien.[4]

In late November, Mal Scurrah was one of several radar operators at RAF Neatishead. They were guiding a flight of Phantom jets engaged in an exercise over the Wash. Suddenly, an unidentified target appeared with no electronic signal. There was measured panic as the RAF failed to identify it. Two Phantoms were vectored onto the target, and one described approaching a huge light. As it closed in, the UFO rocketed upwards in an act that defied description. Scurrah watched his radar screen click off impossible rates of ascent and the UFO was lost above 100,000 feet, far out of reach of the pursuing jet.[5]

The pattern seems unnerving. UFOs were engaged in what resembles surveillance operations focusing on locations with blatant security implications

– the NSA at Menwith Hill, Fylingdales early warning radar, oil rigs and RAF exercises. It is not hard to see why the MoD would take notice.

The main act

But if this was the overture, the main act was soon to follow. During a 48-hour spell just after Christmas, more than a dozen local citizens and over 20 military personnel (including Colonel Charles Halt, the USAF Deputy Base Commander at RAF Bentwaters) were to confront at close quarters what they believed to be a UFO. What did they see? Why was a civil defence alert not mounted? Why did the case not create global headlines? Was this a genuine alien invasion, a threat to East Anglia or disturbing evidence that even the best trained and most highly experienced people can be fooled by an IFO (identified flying object)? If so, this would not be an academic question. The USAF at Bentwaters literally had their fingers on the nuclear button. We should all worry about how properly they acted under the most intense duress imaginable.

Secrets of the pine wood

The huge Forestry Commission land of Rendlesham sits on the south-east Suffolk coast surrounded by small villages and towns such as Woodbridge and Orford. Ipswich is eight miles from the 'landing' site. Deep within this slumbering woodland are a number of facilities whose activities were known mostly by story and legend to the locals. No doubt any hypothetical aliens out there would have found them just as attractive.

From before World War Two, this small area of countryside had been home to state-of-the-art scientific research. Radar was perfected here. Marconi Industries have had a site there. Nuclear weapons detonators were tested on the shore, leading to the sudden evacuation of the village of Shingle Street in a still controversial move (debated in the *Daily Mail* as recently as May 1999). The records on the incident were sealed for 50 years. Such activities have inevitably excited much talk of conspiracy and cover-up.

Initial radar experiments in the forest emerged from the study of pulsed energy beams dreamt up in the 1920s as a deadly weapon. The plan was to find a way to send out a huge burst of energy that would kill off electrical power and cripple an enemy. But this was found to require such a massive output that it was impractical to use, and the fact that lower-powered emissions could show up distant objects as 'echoes' emerged as a vital fringe benefit. If that distant object was an incoming enemy aircraft, a quite unexpected defensive weapon was possible instead of the anticipated offensive one.

The early tests during the 1930s led to much speculation that a 'secret weapon' was being perfected around Rendlesham and the active RAF bases of Woodbridge and Bentwaters. Local tales recalled even today spoke of lights in houses switching on and off, and fishermen returning from trips with skins full of blisters. A standing joke for years locally is that when a car breaks down in the forest, one does not blame a faulty battery, but the 'power outages' caused by these experiments at Bawdsey Manor House (now RAF Bawdsey), where radar was perfected. In the 1960s, the huge nuclear power plant at Sizewell was developed on the rim of the forest. Even the construction workers fell prey to on-going myths. I spoke with one crew worker, Mike Salt, then living on Merseyside. That was in 1978, before the Rendlesham Forest case. Mike told me that he quit his job because of "spooky" happenings at Sizewell.

These incidents lasted months and involved power tools 'firing' up on their own and electric shocks being given by equipment that was supposedly switched off. Odd lights were seen in the deserted plant as well. To the workers the cause was obvious – a ghostly poltergeist. When seen in context with other stories from this area, it is worth suspecting otherwise. Were emitted energy fields from on-going radar experiments generating an intense electric field that by induction 'charged up' equipment at the plant?

A not dissimilar case occurred in 1988 when a new factory opened in Poynton, Cheshire. Staff found a computer screen charging up with an eerie glow even when unplugged. Workers were scared, claiming the building was haunted. Experts called in noted that military radar at the Woodford aeronautics plant was just a quarter mile from the factory. It could 'leak' energy and charge up a faulty screen.[6]

A fascinating UFO case occurred in February 1975 on the beach at Sizewell, less than a mile from the Suffolk nuclear plant. Walking his dog, the local postman encountered a beachball-like object glowing a brilliant fluorescent green-yellow. It floated in from the sea and seemed to electrify the air, creating a static charge that terrified his dog, who fled immediately. The object also left the witness with watering eyes, pounding headache and feeling ill as if suffering from radiation sickness (or so colleagues quipped). He was not able to work for months, and reported how the air around the floating ball smelled pungent – like acid drops – a sign that this energy mass was creating an electro-chemical reaction in the atmosphere.[7]

Several fishermen at Orford claim to have seen these green balls. I spoke with the lighthouse keeper at Orford Ness, who recalled they had appeared under water, too. These UFOs are not alien craft – they are very small, charged energy balls, surely of natural origin. But are they natural phenomena or artificially generated by local experiments?

Between 1949 and 1951, the same phenomenon caused huge consternation to the US government because green fireballs flew above New Mexico near Los Alamos and White Sands – locations where, at the time, the most extensive scientific research on earth was undertaken. Top physicists like Oppenheimer and Teller who built the first atom bomb, rocketry experts such as Werner von Braun who developed Nazi V2 rockets, plus astronomers like Clyde Tombaugh, the man who discovered Pluto, were involved in covert research into these UFOs. Many saw at least one during that period, according to intelligence files made public in 1977. The mystery was never solved, but speculation ranged from a secret spy probe to aliens. No records seem to have suggested that these things might have been a natural by-product of the covert nuclear and weapons research then being undertaken locally.[8]

Cobra Mist

Much of the research carried out post-war in the vicinity of Rendlesham Forest remains shrouded in secrecy. Indeed numerous files remain blocked from view. But half a century of radar development continued in Suffolk. Cobra Mist was an American plan to create what was called an OTH Radar. It ran in the 1970s.

OTH – Over the Horizon – systems use high-energy beams that are projected into space and bounced off the upper fringes of the atmosphere. Unlike radar waves that reflect from low level, basic geometry explains their key advantage. The curvature of the earth limits the range at which a radar signal can be emitted, then strike a target and be reflected back. Once much over the horizon caused by this curvature, no object can be detected. But if the beam passes right through the atmosphere and bounces off the edge of space, the range is increased to hundreds of miles at a stroke via simple angular mathematics. You know the enemy is coming long before they realise your fighter aircraft are being sent in pursuit.

High-energy beams have side effects. They can ionise the surrounding atmosphere, causing communications interference and other potential consequences. Cobra Mist as a project sought to minimise these problems. It is no wonder that there are claims by local people of TV sets and house lights going haywire, or car electrical systems conking out and power tools developing a life of their own. One woman in Laxfield told me of a low-pitched hum she heard in dead of night that gave her nightmares. It was accompanied by TV interference. What seems a collection of unrelated supernatural tales now seems – in light of this research – to be a down-to-earth result of covert scientific tests ongoing in this location.[9]

Cobra Mist ceased operation around the time that the green fireball scared the postman on Sizewell beach. But the NSA became interested – the same American intelligence agency that has massive telecommunications and electronic monitoring operations at Menwith Hill in Yorkshire.

During the 1970s, NSA operations in Yorkshire were associated with strange balls of light seen cavorting about local skies. Some UFOlogists suspected the aliens were keeping tabs on this ultra-secretive agency. Of course, if they were testing experimental radar, these Yorkshire UFOs may have been emerging from the atmosphere as a quite terrestrial side effect.

The code name Cobra Mist conjures up a chill image of more than just a radar beam. It instils visions of a curtain of energy that could strike with deadly force like a cobra, paralysing the enemy. Was this experiment an attempt to turn a defensive radar into an offensive weapon by scrambling electric power? In January 1998, I spoke with one BT telecommunications worker from Woodbridge who in the 1970s had been at the research site on Orford Ness, a coastal spit at the edge of Rendlesham Forest. He knew of Cobra Mist, but insisted that the NSA did not bring a more developed project to Suffolk.

That view is contradicted by certain evidence. A US scientist who worked on Cobra Mist dropped a sheaf of secret documents by accident at Heathrow Airport when he left the country. They discussed the location of the new experiment (to be named Cold Witness). Orford Ness was cited. NSA agents, physicists and MoD scientists from Bawdsey were to run it in strict secrecy. If so, this was a joint UK/USA experiment. The people of Suffolk might not have been top of the list to be advised.

A document also surfaced in 1987 from the Public Record Office in Kew. This was an MoD advisory notice to shipping (circa 1980) that advised against them carrying hazardous cargo close to Orford Ness. No reason was stated. But Colonel Charles Halt, the Base Commander at Bentwaters during the 1980s, told me that there was secret research on the Ness. Sometimes his men were ordered to go in and "clean up the mess" after some sort of incident. They were not allowed to discuss this work.

Ralph Noyes, an Under-Secretary at the MoD, reached the rank of Air Commodore before retiring. He came out in 1983 to support my work on Rendlesham. Noyes told me during many conversations that UFOs were real, the MoD knew this, but also that they were not alien craft. Some form of natural energy occurred within the atmosphere and can potentially be harnessed as a weapon. He was quick to stress this was a personal view, not an MoD stance, but such was his status I suspect the two are not entirely unconnected.

At the time, I knew little of Cobra Mist or local stories that suggested strange forces at work in Suffolk. Given his position at the MoD during the years when Cobra Mist ran, Noyes probably had access to data, but was not free to talk because of the Official Secrets Act. I think he did the next best thing in 1985 by writing a little known but highly intriguing fictional account of UFO encounters in Suffolk. In his tale, he tied local UFOs to a beam weapon that created weird glowing lights over the forest and scrambled electronic systems. It was tested on a coastal spit that he called Blandford Ness.[10]

Disinformation

This important background places the events of Rendlesham Forest in context. Those Christmas 1980 lights may have been coincidental meteors and space junk. But if energy beam experiments were happening close to Rendlesham, a different cause for them has to be considered. Similar green and orange lights have been reported in an area of central Australia around Pine Gap. NASA physicist Dr Richard Haines – unaware of the data in this chapter – told me of his study there when we met in Washington. He suspected they were connected with a covert military experiment ... and the NSA has a facility at Pine Gap.[11]

One can make a strong circumstantial case out of all this for a covert experiment on Orford Ness. The choice of Christmas night may well be relevant. With liberal amounts of spirits flowing, witnesses would be less readily believed. The beam would project mostly out to sea to the north, south and east. Only to the west would it cover land, and there is only sparse habitation amidst forest in that direction. Also, in the middle of the night few people would still be up or too 'merry' to notice. Any story regarding the events that may emerge might not get much Press attention either. Media have limited resources between Christmas and New Year and publish less frequently. A story breaking then could vanish into obscurity.

Moreover, the already predicted burn-up of Cosmos 749 that night offered a terrific smoke-screen. Any glows created as a side effect of an experiment would be assumed to be this space junk. Hardy UFOlogists may consider other options. But tales of a spaceship, complete with aliens, would soon distract them and serve as useful counter-intelligence. UFOlogy would (and did) quickly embrace such claims, and by shouting them from the rooftops rapidly demolished any serious journalism from attaching to the case.

Within 48 hours of the events reported by USAF personnel at Bentwaters, USAF intelligence officers visited RAF Watton, a radar base near Norwich that say they tracked something above Rendlesham Forest. The base reported it to the MoD, but the USAF turned up on December 29 and took away the film "for analysis." When doing so, they told radar staff that a UFO crashed in Rendlesham Forest, aliens floated out and the base commander communicated with them! Why report such an absurd yarn?

I first discovered this allegation from a radar officer at Watton a month later, because, incredibly, nobody was put under pressure not to talk to me. I had no idea that a UFO incident had occurred in the forest. Yet the story passed to me is so wild and unverified it seems crazy. Whilst a UFO encounter seemingly did occur, there is no evidence of a crashed spaceship, alien pilots or the base commander chatting to them. So why were radar staff at Watton told this amazing and presumed lie by USAF intelligence agents?

If the crash had happened as USAF intelligence insist, Watton staff would have no need to know. The radar film would have been taken simply 'on orders' and those present reminded that they had all signed the Official Secrets Act. To be effectively encouraged to talk about it to a UFO buff is suspicious.

Just two weeks before the encounter (mid-December 1980), I addressed senior politicians and top MoD staff at their request in the Houses of Parliament. Was I then singled out for disinformation? It appears that someone wanted news of this case to be perceived by UFOlogy as an absurd alien saga. There was even a clue as to why the crashed alien spaceship theme was chosen. At that time, British cinemas were screening a terrible American movie called *Hangar 18*. It told of how an alien craft crashed, was captured by the US military and taken to a USAF base for storage. Did this give intelligence a suitable idea?

I was not the only UFOlogist to be offered this now discredited version of events. In January 1981, Leiston-based UFO enthusiast Brenda Butler was befriended by a USAF officer from Bentwaters. He claimed to be witness to the whole thing, refused to go public, but freely gave her an account that was much like the one I independently received via Watton that month. On reflection, it seems clear that we were both used. This bizarre tale was what we were expected to believe and offer up to the UFO community. But it was not a true version of the events. As such, one can only presume it was deliberate misinformation.

No USAF witness that I have talked with has ever recognised the real name of Brenda's contact as being a participant in the case. His name features in none of the reports now available. But the man won our trust by providing a secret document not available through the US Freedom of Information Act. This was a letter from a senior USAF officer urging NASA not to accept President Carter's efforts to give funding to launch a UFO study. The Carter project was real, and NASA did reject it. Was this letter a sop given to UFOlogists to make his dubious story of a UFO crash credible? Ex-Base Commander Charles Halt told me in 1997 that the man he suspected as Brenda's source had worked for USAF Public Affairs. If true, this makes it even more interesting that he would choose to feed what we now know to be a ludicrous tale to a UFOlogist.

GULF BREEZE ENCOUNTERS

In November 1987, the north Florida town of Gulf Breeze was gripped by UFO fever. It began when the local newspaper received some photographs taken by a businessman who requested anonymity. Known only as "Mr Ed," there were a large number of the shots, all depicting semi-transparent and highly luminous UFOs hovering in the sky above the town. They resembled reflections of lampshades in a window, a point noted by some critics, but such an origin was emphatically denied.

The pictures were taken by a Polaroid camera, meaning that no negatives were available for analysis. During the next year, the town went UFO crazy, with several further photographs being taken. UFO experts flocked there and sky watches were held by the Bay Bridge, some turning into a circus as media cameras from all over the world joined UFOlogists, New Age mystics and sightseers.

The identity of the initial witness was exposed as Ed Walters, who then began to claim even stranger experiences, including alien contacts. Photographs were taken of a UFO hovering low over a road, and of a beam of light fired at him. But no pictures were provided of the aliens themselves.

Ed Walters was supplied with a sophisticated stereo camera by optical physicist Dr Bruce Maccabee and some further photographs were taken. But these did not depict the same disc-like craft. Visible on these images was a small lit object a foot or so wide.

The fame of Gulf Breeze lasted well into the 1990s, with numerous clips of video – showing orange lights dropping 'satellites' into the bay – taken during the skywatches. Walters wrote two books describing his photographs and alien contacts, the latter showing some similarities with the recently reported alien abductions of Whitley Strieber.

Is it real?
● That something was seen over Gulf Breeze on at least certain occasions seems well established by the film evidence and number of witnesses.
● Whilst never establishing that what was seen were large, alien craft, investigation of the photographs came up with little

Scenarios

For the rest of this chapter, I intend to present the main features of the Rendle-sham story in order. This will cover key civilian witnesses, military personnel who had encounters and the evidence that accompanies their tales. That evidence ranges from physical traces to photos and radiation left behind.

As we progress through this account of the case, we must consider the way in which these events should best be interpreted. Bear in mind the 'party line' fed out by USAF intelligence – and perhaps its public affairs office – via UFOl-ogists like myself. This was that an alien spacecraft crashed in Rendlesham in the early hours of December 26, senior personnel surrounded it and Brigadier General Gordon Williams (overall Wing Commander) communed with small, floating aliens using sign language.

In my mind there is no chance this really happened. The way this news was released has always made me suspect that the truth was more down-to-earth

evidence of trickery.
- Window areas where UFO activity occurs with such frequency are known to exist.

Is it solved?
- After Ed Walters moved house, 'plans' were found that appeared to be designs to build model UFOs like those photographed. Walters argued that they had been planted.
- The shot of the UFO hovering low over the road presented problems when matching the angle at which it hovered with the light splayed out on the ground below.
- Servicemen at the Pensacola Naval Base, across the bay from the skywatch sight in Gulf Breeze, admitted faking sightings to play to the

crowd. They made hot air balloons using candles and plastic bags. Some of the video shows bits of melted plastic dripping off as the 'UFOs' drift over the bay. These are interpreted as 'satellite' UFOs by the expectation of the witnesses.

Conclusion
Hoaxing may – or may not – be involved in the original photographs, although it was clearly a factor in a few subsequent cases once the town became a tourist attraction.

As is often true, witnesses and photographers are not necessarily those doing the hoaxing. The camera can be fooled just as easily as a person. Given the huge media interest and the money on offer to witnesses and UFOlogists, there was

certainly a lot of temptation to play games in Gulf Breeze.

What we can say is that despite all this attention and years of extensive investiga-tion into the evidence, it failed to provide anything close to acceptable scien-tific proof that aliens visited this small town. One is tempted to say that if so many pictures depicting such clear-cut alien craft and sightings made by so many witnesses still failed to establish proof, what case ever could produce a more positive outcome?

Further reading
The Gulf Breeze Sightings. Walters, E. and F. William Morrow, New York, 1989.

and these exaggerations were designed to discredit the whole affair once it got into the public domain. Publicity was inevitable as there were too many witnesses for total containment of the case. Needless to say, Gordon Williams has categorically denied he was involved in any way and is supported by the witnesses from that night, none of whom says they saw him out there.

One alternative scenario is that a covert experiment occurred at the research centre on Orford Ness and that this created glows in the atmosphere as a side effect of its energy beam. Lights seen in this area were a product of that test. Any physiological effects noted by witnesses stemmed from the intense electrical fields generated by the experiment. You can try to fit witness testimony into this possibility as we progress, although it will remain informed speculation because we can never prove such tests did happen.

This case has become such a confusing mess of claim and counter-claim. Some of the tall tales spun about Rendlesham may have been created with that purpose in mind – to bury these experiments beneath a sea of paranoia. Others undoubtedly emerged from uninvolved personnel on the base who were jumping on the bandwagon and making up stories for their own motives, using scraps of information that quickly buzzed around the base at the speed of rumour. Sifting fact from fiction has long dogged this case.

However, these are not the only possible options. There is another scenario that has to be considered as we sort through witnesses and evidence. This was first proposed by Forester Vince Thurkettle in October 1983 when news of the alleged alien landings hit the media and fired-up the sceptics' community to find a solution. Its main proponent has always been astronomer Ian Ridpath, a noted, fair-minded critic regarding UFO reality.[12]

Since 1997, this idea – once believed absurd by UFOlogists (myself included) and still discredited by witnesses – has been superbly reinforced by the skilful investigations of Scottish UFOlogist James Easton. He set out to review the case, anticipating proof of its status as the most important in the UK. But at each turn he found clues that dragged him closer towards the opinion that Ridpath had adopted fourteen years before.

Easton and Ridpath argue that this case may be one of the most glaring examples of misperception in the entire history of UFOs – that every feature can be explained in mundane terms, although often through an unusual combination of circumstances that chanced to come together that weekend. However much you may want to believe that aliens landed in Suffolk, or you may think that the NSA were up to no good on Orford Ness, these ideas may not be necessary. The prospect of this case being no more than a sham has been growing with alarming conviction as new evidence falls into place.

Is it possible that the world's greatest UFO mystery requires nothing more to explain it than a lighthouse, a few rabbits and some stars and meteors? Disturbing as it may seem, the answer has to be 'Yes,' and the in-depth appraisal that follows exposes some telling new evidence that points towards the truth.

Lights in the forest

In the early hours of December 26, quite a few civilians appear to have seen the arrival of strange lights over Rendlesham Forest. It is worth noting that 'lights' are indeed really all that these people describe. None reported a structured alien craft.

Arthur Smekle, a travelling salesman, was returning home to Essex after spending Christmas night with friends at Butley. Passing on the B road near Woodbridge base (some time in the early hours), he saw a "mass of lights" that "looked a bit triangular." This fell into the woods and was lost as he drove by.[13]

At approximately 2.30 am, the Webb family from Martlesham were driving towards Woodbridge when daughter Hayley saw "a bright star" hovering ahead. Roy Webb stopped at the first lay-by and they all had a good look. The light was extremely bright and lit the forest as it shot across the sky at speed.[14]

About this time, Gerry Harris, who ran a small garage near Woodbridge, was locking up for the night when he and his wife saw a huge light over the forest that was illuminating the area. Like Arthur Smekle, he assumed it was just some sort of aircraft connected to the base and went indoors. He only recalled the incident later after the other events of the weekend emerged.[15]

The brilliance of the light was noted by a farmer called Higgins. Brenda Butler and colleague, Dot Street, first heard of him in early 1981 from talk in the local pubs. But it took us until 1984 to find him. This we did by going door-to-door, hunting down all known farmers, tracing a friend who had assisted him that night and eventually discovering his new whereabouts. Higgins had moved 200 miles from Suffolk. The brilliant light on December 26 scared his cows, causing them to flee onto a road. A passing taxi struck and injured some. He assumed the light was an aircraft, although he heard no sound.

Higgins complained to base public affairs and was told that no air traffic was flying that night. A week or two later, he heard the rumours filling the villages about UFOs. Higgins went back to the air base and told public affairs that if they had no aircraft up that night, it must have been the UFO that scared his cows. But as the USAF were supposed to be protecting Suffolk, it was still their fault! Soon after this second call, Higgins sold up and relocated. All he would tell us when we found him was that he was paid compensation by the USAF and that helped him to move. The USAF public affairs office denied to me that there was any truth to this story.

From these accounts, we can see that a very bright light did appear over Rendlesham between 2 and 3 am on December 26. It was seen to hover, glide and rush away at speed, but made no sound. A vague triangular shape was recounted, but in essence it was just a light. It scared cows well used to seeing aircraft.

At 2.50 am, a very bright fireball meteor is known from the records of the British Astronomical Association to have crossed the sky. It was in view about two or three seconds. Was this what was seen by Arthur Smekle? Did it scare Farmer Higgins' cows?

The Webb family gave the time of their sighting as 20 minutes earlier than the certain appearance of the meteor and saw the light for longer than a couple of seconds. They had time to see it, drive to a halt in a lay-by and still observe the phenomenon before it streaked away. It seems highly unlikely that the Webb family saw the meteor, although it is not improbable that other witnesses, such as Arthur Smekle, did.

East Gate

The main source of the action shortly before 3 am on December 26 was just off the B road that Arthur Smekle had been travelling. It was at the East Gate complex of the Woodbridge Air Base, linked to Bentwaters across two miles of forest. Both Bentwaters and Woodbridge, its daughter base, were entirely staffed by the USAF, which leased ex-wartime RAF land from the MoD as part of their NATO defensive duties. They moved out in the mid-1990s after the end of the Cold War. This vast pine wood is also now very different from 1980. It has lost thousands of trees following a devastating hurricane.

On December 26, a two-man patrol from the US security police law enforcement unit was driving near the East Gate around 3 am. The supervisor was Staff Sergeant Budd Steffans. With him was Airman First Class John Burroughs. Steffans has not given any public comment on the case. Burroughs is another matter. He remained with the USAF until 1988, when he retired with the rank of sergeant. He signed a report form a few days after the incident on behalf of the Deputy Commander at Bentwaters (Colonel Charles Halt). This was retrieved by James Easton in late 1997 from an American UFO researcher who had held this document along with other signed USAF statements about the case. These had evidently been supplied some time after 1985. The forms contain written comments by USAF Base Commander Charles Halt regarding his views on the witnesses, possibly a reason for their lack of public disclosure.

I met Burroughs in Arizona, in 1989. He gave his first public interview in 1991, and in 1994 agreed to a lengthy interview for ITV's *Strange But True?* documentary that I put together with London Weekend Television. These are the main sources of data used for this part of the story (dates in brackets indicate which source). The accounts are consistent, although the written version by Burroughs signed approximately January 2, 1981, is a little muted compared with verbal testimony. This I do not find surprising as Burroughs told me he had not reported everything to the USAF in his statement. Some parts were just too amazing to be believed.

Steffans first saw a light in the sky falling towards the forest. The time may have been about 2.50 am. Nobody is certain, but the base logs indicate 3 am for the first call to security control. As such, it is more than possible that Steffans saw the fireball (assumed to be a bright meteor) just before this time.

Burroughs followed the gaze of his watch supervisor and says (1994): "At first, I didn't see anything. But then I did." This suggests that the initial brilliant light was no longer visible, supporting the theory of a very short-lived meteor. When Burroughs and Steffans both saw something in the trees, it was different, just lights (1981), "red and blue – the red one above the blue one, and they were flashing on and off." At the time they were looking along a small road leading due east from the Woodbridge gate towards the forest. There was a line of pine trees several hundred yards ahead and many trees to the south.

I first stood in this spot in winter 1983, when the view was still much the same. Although looking directly towards Orford Ness, the four miles and countless trees in between made ground lights on the spit invisible. There were, in fact, a bank of lights on Orford, part of the research site. Also, the bright beam of the lighthouse was just to its right, but you cannot see these lights from the gate.

The Shipwash lightship – several miles south of Orford, guarding the river entrance towards Felixstowe – was invisible through thick pines surrounding the Forestry Commission offices in Tangham Woods. However, what one could see – very clearly – from the gate was the beam from the Orford Ness light-house. This swept in an approximate five-second cycle, creating what looked like a bright white glow above the tree tops straight ahead from where Burroughs and Steffans now sat in their jeeps. As the rest of the area is dark forest and the base lights are to the rear, it was impossible to miss this beam from this spot.

Hunting the lights

Burroughs got out of his cruiser, opened the gate and both men drove to the spot where a small, rutted, logging track led east into the forest. They stopped here and dismounted for a closer look at the lights. From this spot in 1983 I saw the lighthouse beam as more prominent and it was now possible to glimpse the pulsing light through trees ahead. But I never saw other lights from here. The odd aspect about the pulsing white lighthouse from the logging track was that it appeared at ground level. One is four miles from Orford Ness. Land slopes gently towards the coast, placing the light – line of sight – on the horizon. Owing to trees, I could not yet see the Shipwash light-ship, off to the right.

From their new vantage point, Burroughs described the anomaly (1981) as "a white light shining onto the trees and [I] could still see the red and blue lights. We decided we better go call it in." He elaborated (1994): "I just stared forward into the trees. I was scared … and had this very weird feeling." Burroughs likened the sight to "Christmas tree lights," and noted how the colours were beneath the white beam. This glow suddenly brightened (he thought it was flying at them), and both men chose to quit the logging track.

Standing at that spot, I saw just this effect as the rotating Orford Ness beacon shone straight at me. It did temporarily 'flare up' and seem closer, then diminish until in line of sight on its next cycle. Such a description of the white beam is consistent with the lighthouse. Indeed, it is difficult to imagine how these men could not have seen that rotating beacon from where they were then standing. But they do not refer to seeing a strange light in relation to this land-mark. Here, Burroughs never mentions seeing the lighthouse at all. Does that prove that the white light they were staring at was simply this beacon?

There is a problem, aside from the presence of coloured lights not emitted by the lighthouse. John Burroughs told me in 1989 (and repeated in 1994) that he was familiar with the forest. He had been on base for eighteen months; he picnicked in the woods; he often saw the lighthouse. Moreover, he had patrolled the East Gate before. From here, one can hardly miss seeing the light-house beam sweep over trees. It is not known how familiar Steffans was with the area, but Burroughs was adamant about his familiarity. Indeed, if all that Burroughs saw was the lighthouse, then why on that specific night did it cause such panic?

Now back at the East Gate, Burroughs called on the radio phone to CSC (Centre for Security Control). He had trouble persuading the law enforcement chief that his sighting was not a joke given that it was Christmas night. As he radioed through, Burroughs says (1981): "The whole time I could see the lights. The white light was almost at the edge of the road, and the blue and red lights were still out in the woods." I remind readers that from the gate only the white beam above the tree tops was visible to me.

After convincing the CSC that the sighting was real, a call was put out by the base to a member of the 81st Tactical Fighter Wing security patrol. Burroughs and Steffans were USAF police in the sense of law enforcement (seeking out misdemeanours). The Security Police would respond to any possible threats to the base and were then on alert for IRA intrusions. The security patrol was told by radio to respond to a "non-emergency of critical importance." This means the investigating officer was instructed to drive at speed using lights and sirens to clear traffic as he crossed roads. A possible security breach was imminent.

Help arrives

The man sent to investigate was Sergeant Jim Penniston. He had been on base only six months, but in Britain since 1975. His original posting was to RAF Alconbury from where, according to reports years later in *Jane's Defence Weekly*, US Stealth fighters were being flown in top secret. Penniston was a veteran of 30 air crashes – seeking possible security implications of these. From the moment he reached the East Gate, this was what he thought he was dealing with on December 26.

Penniston has spoken often. In 1983, he gave several brief interviews (insisting we only call him Jim) whilst Brenda Butler, Dot Street and I were writing our first report on the case.[16] He convinced us then as the most persuasive witness we encountered, but was clearly very worried about what he was doing. Making it plain that he was afraid of surveillance, it was only years later that he explained why. Penniston says he was debriefed by both British and American intelligence agencies a week or two after the incident, and even found a listening device planted in his room when reassigned to a base in Indiana. Penniston remained at Bentwaters until 1984, leaving the USAF just before his first fully on-the-record interview in 1994.

Other interview sources for Penniston are *Strange But True?* in 1994. As with Burroughs, the complete interview transcripts that I made at the time are far more extensive than extracts we could use on TV. Penniston also gave an interview in 1997 to US science reporter Sally Rayl. His original statement to the base was signed in January 1981, and has also now been secured by James Easton.

Penniston was not on his own when he arrived at the East Gate to meet Burroughs and Steffans. He was driven there by Airman First Class Ed Cabansag, also from his security patrol unit. Cabansag filled out a statement in 1981. We have it. But so far, like Steffans, he has given no interviews. It has been a common pattern that the Rendlesham USAF witnesses all stayed silent until they left the service years later. Cabansag's 1981 statement is rather muted. In a note written on the side, Halt suggests that it is a "cleaned up" version, and that the young airman was "shook up to the point that he did not want to talk." He understood that the airman "still worries today" (it would be about 1985 when Halt penned this note).

This is important because it suggests that the signed statements logged with the base by the witnesses in January 1981 may have been deliberately watered down. Burroughs has long said this. Similarly, Penniston consistently claimed in verbal interviews that he restricted the statement given to Halt in 1981. He claims that the witnesses agreed this protocol between them before leaving the forest, fearing the consequences on their careers. "John and I decided that we could not tell them everything that had happened. It was too fantastic ... we just told (the base) we had seen some lights in the woods ..." (1994). This claim needs to be considered because the statement made by Penniston in 1981 is massively at odds with his verbal accounts on subsequent occasions (even when he spoke as just 'Jim' back in 1983).

There are only two obvious ways to interpret this difference. Whichever one you follow significantly affects how you interpret the case. You may, as Ridpath infers, assume that the original statement was basically correct and all that was seen by these men were lights. Penniston's later accounts would then be regarded as confabulation. Alternatively, you accept his claims, made years before his written statement surfaced, that his verbal account is the true version and the 1981 statement was deliberately muted.

Tension mounts

After arrival at the East Gate with Cabansag, Penniston saw the lights immediately. He says (1994): "They appeared to be a fire, the sort of thing you might get from chemicals burning after an air crash. Based on my experience, that is what I assumed it was." Oddly, in 1997 Penniston times the incident at 12.02 am – three hours earlier than any other witness or the base records. He claims, "I remember that distinctly," adding, "I gazed about 300 metres into the distance and saw ... orange, red and bluish types of glowing light, pretty standard with aircraft crashes." Compare this with his 1981 statement where "A large yellow glowing light was emitting above the trees. In the centre of the lighted area directly in the centre ground level there was a red light blinking on and off at 5- to 10-second intervals." This is not the same thing.

These discrepancies show the confusion. The white light over the trees as seen from the East Gate surely has to be the lighthouse beam. What then of Burroughs' "Christmas tree" lights or the chemical fire Penniston now reports? Cabansag never even mentions seeing any lights from the gate area in his 1981 report.

Penniston says he tried hard to persuade Steffans and Burroughs that they saw a plane crash. Both refused to accept this. Steffans insisted the light initially landed in a controlled manner. "I discounted that," says Penniston (1997) because "the dense woodland made any controlled landing improbable" (1994). Even so, he called CSC and began the air crash check procedures. In this regard, it is interesting that in 1998 I received an anonymous letter after a lecture on the case at Woodbridge Town Hall. It claimed that a senior officer from the Suffolk Fire Brigade went to the forest on request of the base to assess the situation. If so, he has never spoken publicly on the matter and the fire service officially deny the tale. However, standard procedures that Penniston says he initiated would include the calling of the fire brigade to the scene.

Penniston says his doubts about a plane crash grew when he saw the reactions of Burroughs and Steffans. Both had seen air crashes and were trained to respond, but they were reacting as if in total shock. At one point, the sergeant walked on his own down the road, stared at the lights and returned to the gate phone. On the phone, Staff Sergeant Coffey – the CSC controller on duty – had asked Penniston to check if some sort of flare or marker lights might be in the woods. He was soon joined on the phone by the CSC chief, Fred Buran, who had been alerted to a possible security threat. He arrived (as his 1981 statement records) more or less as Penniston was ruling out these flares and marker lights ideas, and saying that what he saw was like nothing he had witnessed before in this area. It may be significant that the colours described (orange and blue) are commonly created as an optical illusion when one's eyes stare at a bright white light.

By now, Master Sergeant J. D. Chandler, the on-duty flight chief, was also involved in the conversations back on base. Events were becoming disconcerting. Chandler had monitored radio traffic and decided he must act. The men debated the possibility of radar confirmation. This is never mentioned in the 1981 base reports, but Penniston says (1994), "I asked the control to check for any radar returns, but was not hopeful."

In 1997, he added: "Chandler asked me to stand by while he contacted the control tower at Bentwaters and also at Woodbridge ... they in turn were in contact with Heathrow Airport in London, RAF Bawdsey and Eastern Radar in East Anglia." Bawdsey – the radar research unit close to Woodbridge – supposedly triangulated various radar sightings. About fifteen minutes earlier, radar had spotted an object three miles out, but lost contact over Rendlesham. Eastern Radar is another name for RAF Watton, the base from where the recordings were taken by USAF intelligence and from whom I first heard of the Rendlesham encounter.

It was now decided that the only course was to go into the forest and investigate. Penniston was given permission to take two men with him (one from law enforcement and one from security). They had to leave their weapons behind as they were not allowed to enter British territory still carrying them. Steffans stayed at the gate with these and Chandler drove out to the gate to secure them. Penniston took Burroughs and Cabansag along the East Gate road and drove to the edge of the logging track into the forest. After a short distance, its frozen, rutted surface made progress hazardous and they left their vehicles behind and set off on foot in pursuit of the lights that had now been in view for 20 minutes. To do this, they had to walk deep into the forest in the dark. It was inevitable that they would become disorientated in this process.

The close encounter

Burroughs says of this forest trek (1981), "The lights were moving back and they appeared to stop in amongst a bunch of trees." Penniston (1994) adds that "part of the woods was brightly lit by a mass of fluctuating light ... inside the tree line, it appeared larger than we expected." But he still thought it was a blazing aircraft. By now, communications back to CSC via Steffans at the gate were suffering problems. Penniston (1997) says: "Our radios – the standard

Motorola type – were experiencing a dampening effect common with atmospheric conditions. Basically they were breaking up."

In 1994 and 1997, Penniston claimed he decided to ask Cabansag to remain on the logging track as a further relay point. This appears not to be true. Penniston may have been trying to suppress the fact that Cabansag was a full-scale witness. This man evidently wishes not to talk about the night.

Burroughs indicated (1994) that all three of them (he, Penniston and Cabansag) proceeded on foot to try to close in on the lights that had now receded back into the trees ahead. This behaviour is, of course, quite consistent with something several miles distant and not as close as these men evidently assumed it was. You sense you are closing in, but never quite seem to get right up to it. The lighthouse and other lights on Orford Ness were still over three miles ahead of them and could easily fit this pattern.

In his 1981 report, Cabansag recalls how he headed off with the others into the trees: "While we walked, each of us would see the lights – blue, red, white and yellow … We would see them periodically, but not in a specific pattern … We advised CSC and proceeded in extreme caution."

Cabansag also notes that at this point Chandler joined them from the East Gate and came onto the logging track nearby. He says, "CSC was not reading our transmissions very well." So Chandler – not Cabansag – remained by the jeep as the new relay point whilst the three men went deeper into the woods, heading for a small clearing where they believed the source of the lights to be. Chandler's 1981 statement confirms this: "When I arrived, Penniston, Burroughs and Cabansag had entered the wooded area just beyond … We set up a radio relay … Penniston [now] relayed that he was close enough to the object to determine that it was a definite mechanical object."

Buran, at CSC back on base, supports this. His 1981 form reports his monitoring of the relayed radio traffic and how "They appeared to get very close to the lights, and at one point Sgt Penniston stated that it was a definite metallic object."

In his 1981 statement, Burroughs said little of this close encounter phase. He noted "as we were coming into the trees there were strange noises, like a woman screaming. Also the woods lit up and you could hear the farm animals making a lot of noise … All three of us hit the ground and whatever it was started moving back towards the open field. After a minute or two, we got up and moved into the open field." The animals may have been cattle owned by Higgins hit by a taxi around this time seemingly less than half a mile from where the airmen now were. The 'scream' is almost certainly a local bird. I

185

have heard it at night in the forest. Its call is very shrill and does indeed resemble a screaming woman.

On the far side of the small field was a house, usually termed a 'farmhouse' by USAF reports. In fact, the 'farmhouse' was an isolated home surrounded by forest. This house was owned by gamekeeper Vic Boast. I met him and his family in January 1984. He denied any knowledge of the case, but looked edgy. His young daughter was far more animated. She was talking to me eagerly about how "it" was "so big" that she could not understand how it fitted into the forest. But before I could get any elaboration out of her about this curious remark, her father pulled the child away and sent Brenda, Dot and me packing.

BELGIAN WAFFLE

On the night of November 29, 1989, a major wave of sightings struck Belgium. They spread over much of the country, but many sightings occurred near Eupen.

Police officers formed a number of witnesses as the events occurred late at night. The object was described as a vast triangle "the size of a football field" with many bright lights and moving extremely slowly. A faint buzzing noise was heard by some witnesses.

Considerable media publicity followed and dozens more sightings were collated during following weeks. Often, these involved swooping white circles of light in some form of triangle.

Initial suspicions fell upon the activities of the USAF flying Stealth aircraft, but the US government denied any flights over friendly territory. In March 1990, concern amongst the Belgian government was such that

they agreed a policy with UFO group SOBEPS. Aircraft would be placed on standby and SOBEPS should monitor cases that the public would be encouraged to report. If anything was seen, the aircraft would be sent in pursuit. These experiments proved fruitless because no triangles appeared.

However, at 2 am on March 30, police officers at Wavre reported seeing some lights. Two Belgian Air Force jets were scrambled, but saw nothing. A photograph was subsequently submitted of a set of lights seen 100 miles away later in the night. Study of the two air force jet airborne telemetry readings seemed to indicate that unidentified radar tracks had been recorded during their forlorn chase. The case made world headlines when the Belgian Air Force called a Press conference and made the film of the airborne radar available.

Is it real?

● There is no question that various objects were seen in the sky as dozens of reliable witnesses described similar things, plus there was film and radar support during this four-month period.

● Some of the data was compiled during a unique skywatch exercise where UFOlogists and government sources combined. Before holding the Press conference, some sources for the UFO activity were eliminated, but the Air Force were forced into premature revelations.

● The basic triangle shape matched the type of UFO then becoming common all over Europe.

Is it solved?

● In no case was there a direct correlation between visual sightings, radar images and film. In fact, there was a haphazard collection of phenomena that only superficially

The Boasts have constantly maintained a discreet silence on the matter, occasionally denying they made any sightings, although seeming to support the view that something did occur. However, at the time they owned several fierce guard dogs. These did indeed make "an awful lot of noise" whenever someone crossed the small field at night as the three airmen did in 1980. This may be part of the cause for what the USAF men report as "farmyard animals" going crazy when near this clearing.

Penniston's 1981 account tells how the light illuminated an area of 30 metres dead ahead. That means they were looking straight at the Boast house and a few miles beyond this towards the Orford Ness lighthouse. From this

appear to slot together.

- The slow-moving, football field-size object that began the wave in November 1989 has been reported elsewhere in Europe around the same time (see p 86). The UK cases were explained as a mid-air refuelling exercise involving tankers and jets flying in high formation, strung up with unusually bright lights to avoid collisions. The slow motion and faint noises were due to many aircraft at great height. The exercises were part of NATO practices for long-range bombing missions. It is possible these occurred over Belgium, too.
- The swooping white lights in a triangle were almost certainly laser light displays, then novel and unfamiliar to most people. When reflected off low cloud, they produce just such an effect.
- The lights seen by the police at Wavre matched the position of stars

distorted by weather conditions. Full assessment of radar images later revealed the probability that readings from the military jet were spurious and connected to echoes of the companion aircraft. The photograph was not directly connected to the other encounters. It coincided with the flight of a commercial jet through the area.

Conclusion

Although some UFOlogists regard this wave as one of the most important on record, it does appear likely to be resolved. This seems to be a case where a number of unrelated answers combine to produce what looks like a collected flap of sightings.

In truth, the publicity for the first event (a mid-air refuelling exercise?) led to many witnesses being encouraged to report what they later saw (then new and strange-looking laser lights, probably

various other things, too). This resulted in the skywatch exercise that 'made' the case, although its three components – visual, radar and photograph – are separate in time and space, and clearly not a single phenomenon. None stand up to much scrutiny as evidence for real UFOs.

Each component seems in retrospect to be suspect. The outcome is a complex wave with many parts, but each relatively simple to explain. The glue that holds them together was media publicity and UFO enthusiast support that labelled this a wave and escalated its significance on that faulty premise.

Further reading
Article by Paul Vanbrabant. *New UFOlogist*, 1, 1992.

clearing, one certainly could see not just the sweeping beam, but the rotating spotlight as well through trees and the coloured lights on buildings ahead. In addition, from here the Shipwash lightship was visible off to the south (right when facing the coast). This was notable if one looked for it, but was much less prominent than Orford Ness. However, if unaware that there was both a lighthouse and a lightship visible from this position in thick woods, one might have been easily confused.

The lightship flashed in the distance as one might expect a lighthouse to do. It was not a prominent light and well away from the UFOs, meaning one would be unlikely to refer to it in connection with the case. If one mistakenly assumed that it was the Orford Ness lighthouse, what would one make of the real lighthouse, which was far brighter, seemingly at ground level through trees, a lot closer than the lightship and dead ahead? Knowing its origin, I was never in any doubt. Had I not known what this bright light in front of me was on my first nocturnal trip to the woods I do wonder. Being inside a forest with no coastal features in evidence, strangers may not first contemplate this explanation for a pulsating light.

Penniston's 1981 account describes what the men now saw: "The object was producing red and blue light. The blue light was steady and projecting under the object. It was lighting up an area directly under and extending a metre or two out." Could this be the lighthouse with its sweeping beam? If so, I never saw it display any colours. It was just a small white light. Penniston insists that the thing they saw could not possibly be a lighthouse. It was his dramatic observation (made from 50 metres according to his 1981 statement) that led to his radio report that it was "definitely mechanical."

Cabansag's 1981 version says that inside the forest they did not see blue and red lights (to my recall, the building lights were less easy to see in trees than from the clearing). He adds: "Only the beacon light was still blinking ... we could see a glowing near the beacon light, but as we got closer we found it to be a lit-up farmhouse [the Boast house] ... We thought it had to be an aircraft accident."

This version by the young airman seems consistent with seeing just the Orford Ness lighthouse. The blinking beacon was directly behind the Boast house, as indeed was Orford Ness itself.

The nightmare

If these reports were all we had to go on it would be very difficult to deny the facts. These men were staring straight at the Orford Ness lighthouse describing a light largely consistent with it. Never do they relate the UFO to the lighthouse in the same written or verbal testimony, although if both were visible at once then such an omission seems hard to explain as the two should have been almost side by side.

The coloured lights are certainly harder to account for. They might be an optical illusion, but the lighthouse never seemed bright enough to me to affect the retina. Do they suggest some nefarious activity at the NSA site? The lights were obviously just distant lights on a building to me (and I did not at first know what they were). Nor were the building lights visible from places where much of the early sighting occurred. The huge bulk of the trees rearing up in front simply blocked them out. Even so, on balance it is hard not to conclude that Orford Ness beacon was the principal yellow-white UFO.

However, this has to be judged against the verbal testimony of Burroughs and Penniston. This they claim to be the real version of what happened during the close encounter phase. They are either making this story up – and neither of them strike me as liars – or are describing events simply not consistent with the beacon light. Cabansag's confirmation or denial of this new account would be of great value, but his report form merely ignores it, as, of course, do the forms of other (as yet publicly silent) witnesses.

Penniston (1994) says of his close approach: "When I got within 100 feet of the suspected aircraft crash, what appeared to be five or six multi-coloured lights like a flame appeared … A definite form – some sort of image – was emerging from the glow. I knew at that moment that this was not an air crash."

Burroughs (1994) confirms that the "lights seemed to get brighter and more defined. There was more colour behind them. The best way that I can describe the object that emerged from this background is that it was a very brilliant white light with multi-coloured lights inside." He added that it was hard really to call it an object, although it had a vaguely triangular shape. In 1989 he told me, "Words are useless in this context." Burroughs stated that the shock of this sudden confrontation caused all three men to fall on the ground. "We ate dirt," he reported (1994). In fact, Burroughs also told me that he felt his memory of what occurred during the close encounter phase was distorted. Time lost its meaning. He was unsure how long this all lasted.

Penniston (1997) says that the light was so bright they had to squint to look at it. Is that a clue suggesting the orange and blue colours were optical effects in the eye caused by such brilliance? The thing was above the clearing and had frightened local wildlife, such as rabbits and cows, which were running about in a frenzy. Of course, these airmen might have scared them, too, just as any sound did with the Boasts' dogs. Penniston adds that Burroughs was in a curious state and "did not acknowledge what I was saying." Does this support the view of the other airman that he was in some altered state of consciousness?

The craft that Penniston now saw was (1994) "the size of a tank and triangular. It had a very smooth surface – almost like glass ... There were no sharp edges ... The fabric of the craft was moulded like black glass, but opaque or misty. It moved from the black fabric of the craft into the coloured lights imperceptibly. It just blended together." In 1983, Penniston had described it as "dirty" and "off white," different but interestingly similar. Both define a very unusual 'amorphous' UFO as seen by Burroughs.

In 1994, Penniston refused to discuss whether photographs were taken, although Burroughs (1989) implied they were. In 1983, Penniston stated that they were taken. He was more adamant in 1997, and said that he took all 36 pictures on his roll of film at very close quarters. He even sketched the object (see illustrations). What happened to the film? The unprocessed images were taken to the base laboratory next day. He never saw them again. The laboratory simply advised that the pictures did not come out. This Penniston finds hard to understand as the camera and film were of high quality and he was standing close to a brightly lit object.

Both he and Burroughs do agree on one aspect of the close encounter segment not mentioned in their 1981 report – that the air was alive with some kind of energy. This seems to have distorted the very nature of time and space. Burroughs (1994) stated: "The nearer we got to that thing, the more uneasy I felt. This was more than just apprehension. It felt almost like static electricity in the air and was terribly uncomfortable ... It was as if I was moving in slow motion. I felt really hot and the hair was standing up on the back of my head." He was sure that reality itself was being distorted by close proximity to this light source.

Penniston (1994) adds: "The air was filled with electricity – like static. You could feel it on your skin as we approached the object. There was also a sense of slowness, like time itself was an effort." He had to crawl towards the object like wading through treacle. Everything was distorted. He added in 1997 that "it was eerily quiet."

These comments are classic descriptions of what UFOlogy terms "the Oz Factor" and suggest exposure to an electric field. Many close encounter witnesses report that perception and reality are distorted near to a UFO. Evidence suggests that witnesses exit normal consciousness into an altered state.[17]

Burroughs seems to have been 'out of it' at this point. In 1989, he discussed with me that he might have suffered a time lapse and considered regression hypnosis. I sensed he was wary of probing too deeply, but felt there was more to his sighting that he could not bring to mind. He told me one other witness did have a time lapse. We now know it was Penniston, who maintains he went right up to the UFO and actually touched it, one of the few people ever to make such a claim. He says the surface felt smooth and glassy. But he added that there was an area the size of a car number plate which was etched or engraved with symbols.

In 1994, Penniston likened the symbols to mirror writing: they were not a foreign language and seemed vaguely familiar. He has reproduced these images. He also underwent regression conducted by a psychologist in September and November 1994, just after his *Strange But True?* interview. The sessions were at the suggestion of someone who wanted to write a book with him. Penniston says he was wary because of the risk of data contamination and his doubts regarding the credibility of hypnosis, but there was clear evidence in his conscious recall of a memory jump. He had "backed away from the craft because the light was starting to get brighter ... The next thing I knew, I was standing about 20 feet away from the craft with Burroughs." He had no idea how this 'lapse' occurred.

The first hypnosis session brought little new. The second was bizarre. In this, he seemingly 'channelled' data in his mind from entities behind the UFO. However, Penniston does not report them as aliens. He says they were humans from the far future coming back in time to obtain DNA material. This is because the future is a dark, polluted hell where human reproduction has become very difficult. Abductees report alien contact, but are, in fact, contacting themselves.

Psychologists will no doubt find wonderful analogies here as if Penniston's subconscious was trying to tell him something in a symbolic fashion about the origin of these ideas! Penniston says he is unsure how to interpret these hypnotic images. "I don't know if this information is real in any sense, if it has been planted in my mind or if any of it is actually rooted in truth as we know it." (1997)

The departure

In 1994, Penniston reported how the object left the forest. It retracted three beams of light from the base and climbed slowly, edging back through trees at walking pace. Then it "shot away at extreme speed – faster than any aircraft I have ever seen."

Burroughs (1994) said: "One minute it was there. Then it climbed skyward and was gone like a blur. There was no sound and I felt no blast of air from any exhausts." They were later told that a witness in the base control tower saw a white light streaking skywards about this time. He has not come forward.

The men left the forest, allegedly disorientated and – if the hypnosis is to be believed – lost by the base for at least 30 minutes. Burroughs also seems to feel that they may have been out of contact with CSC for some time, but is uncertain on this point. The written statements by all of the men – including CSC control – do not support that. There seems no way they are compatible with any substantial time lapse.

Coda

Burroughs adds little after the departure of the UFO. He did report (1994) that they wandered about the forest for an unknown time rather disorientated until found by others emerging from the trees.

Penniston is more specific. He says that after the UFO left, they saw the same coloured lights out towards the coast. He notes (1997): "We only got about 300 yards into the woods before we turned around. We still had no radio contact … we were not even getting squelch. We went back to the clearing."

Burroughs now found indentations on the ground near where the close encounter happened. They made a note of the position and set off back towards the East Gate. In 1997, Penniston claimed the base had lost contact with them for three hours, but he judged the start time of the encounter at midnight – far too early.

From the 1981 statements, Chandler reports that Penniston's radio relays kept advising how the UFO was now even further away. After some time (not specified) Penniston "arrived at a beacon light," but this was not the same as the initial UFO. Chandler saw nothing during his time on the logging track – even though all the other witnesses had seen lights from here and the Orford Ness lighthouse should have been visible. Does this mean that Chandler recognised the lighthouse for what it was?

Buran notes that when Penniston reported closest contact with the object "all of a sudden [he] said they had gone past it and were looking at a marker beacon that was in the same general direction as the other lights." When Buran reasonably thus asked if this beacon had therefore been the cause of all the fuss, Buran states Penniston was adamant that "Had I seen the other lights, I would know the difference. Sgt Penniston seemed agitated at this point." At 3.54 am, Buran says he ordered the men back to base.

In Penniston's version of events, it was he who decided when to return and, of course, he was 'lost' in the woods for some time as yet. It is possible to estimate how long after 3 am (from the initial call to CSC) it would have taken for all of these events to take place. Thirty to forty minutes seem reasonable. If Buran's timing of the end of the mission is accurate, there is little leeway for more than a few minutes' time lapse at most. Certainly nothing like three hours. Even half an hour would be a stretch.

Furthermore, if the accounts by Chandler and Buran are accurate (and neither man has given verbal interviews), there is no suggestion of any loss of radio contact, just hints of distorted reception. Nor were Chandler and Buran seemingly aware that the airmen were undergoing an amazing close encounter with a glassy-surfaced object that was ripping apart the fabric of space and time with an intense electrical field. Indeed, Chandler stood only a few hundred feet away from these men, but did not report anything.

How is all of this possible? It is worth noting that in UFO close encounters it is common for witnesses in an altered state who are very close to a UFO to experience things whilst unaffected people relatively nearby see nothing. This clue demonstrates to most serious UFOlogists the subjective nature of close encounters. But surely the CSC and Chandler should have realised that something had gone wrong.

Another witness who tells of the aftermath is Geraldo Valdes-Sanchez, a member of B flight with the base security that night. He contacted James Easton in 1998 to give his version of events. Geraldo reports that he had been on base four months. He knew Burroughs and Penniston well and is adamant they are telling the truth. He says: "It was a bitterly cold night and clear ... radio communications were disrupted on and off due to some type of atmospheric disturbance ... We switched channels all night. Contact had been lost with [the men]." This again disputes the 1981 report form statements.

Because of the security threat imposed by the incident, Geraldo says that he was part of a team sent to the gate to investigate. They found the phone off the hook. Weapons were also missing. He adds: "I could clearly see the lights from the gate, just outside the [East Gate]. It was next to the road. They were intermittent lights, very bright, 15–20 feet above the ground. They were pulsating, and from what I recall there were three lights, red, green and blue. It made no noise, but it defied gravity. It was really weird and scary." He continues:

"Burroughs and Penniston finally showed up. I somehow think something happened to them. They were acting strange. We were then told to go back to our posts." Next day people in laboratory coats were at the site in the woods examining the damage.

Many individuals have come forward since the case gained so much notoriety saying that they were involved in some way. We can only listen to what they have to say and assess its context. Whilst some may be genuine witnesses to peripheral features of the case, others are certainly cashing in or spinning a line. It is hard to distinguish. Any attempt to understand the events of December 26 also depends on whether you work from the verbal testimony of Burroughs and Penniston or the recently traced 1981 statements.

These strongly support the theory that the men were fooled by the lighthouse. Indeed, Burroughs and Cabansag add at the end of their written reports a critical fact. They pursued a light for two miles out towards the coast after the close encounter phase of the sighting had ended. Burroughs notes that it was only after this long chase that "we could see it was coming from a lighthouse." Cabansag adds, "We ran and walked a good two miles past our vehicle, until we got to a vantage point where we could determine that what we were chasing was only a beacon light off in the distance." He even notes that "the beacon light turned out to be the yellow light" they had followed all the way into the woods from the logging road.

There is no question that this sequel describes the Orford Ness lighthouse and that if these statements are in any way accurate, it played a vital part in the whole sighting. Another factor needs to be borne in mind. These statements now include a lengthy new part of the story during which the men walked two miles out towards Orford Ness in pursuit of a glow in the forest, eventually recognised it as the lighthouse beacon and then returned. There is little prospect that this could have all been concluded between approximately 3.10 and 3.54 (the time the men were in the forest according to the 1981 statements). A four-mile trek through this area in the dark would surely take an hour by itself. Buran may have called them back at 3.54, but they evidently must have been away for longer than that and perhaps were even missing for some time afterwards. Comments on any alleged time lapse need to bear this confusion of times in mind.

Interpretation

These revelations about the lighthouse chase are stunning. In their frequent interviews, Burroughs and Penniston have never volunteered any story regarding the pursuit of a light that proved to be Orford Ness lighthouse. Indeed, according to Cabansag the yellow light seen by these men all along proved to be precisely that. Penniston's 1981 report does make very clear that the main sighting was different from this beacon. Moreover, one assumes that once having realised their folly and identified the second light source for what it was, they might have accepted they had been fooled and quit while ahead. Instead, they filed official reports, allowed these to be logged with the MoD and Pentagon, were debriefed by intelligence officers and remain to this day adamant the lighthouse was not the primary cause.

Why was this? Were the men now in too deep to admit a mistake? Did they feel they had to justify the use of military resources chasing a light? This seems unlikely, as they were reacting as any trained personnel on alert for an attack should do. On balance, it does seem that even after the mistaken pursuit of Orford Ness the airmen sincerely believed they had initially experienced something stranger. However, the very fact that they did not recognise Orford Ness lighthouse inside the forest is highly significant as is their failure to mention this event during all those interviews, especially given John Burroughs' insistence on how familiar he was with the woods. Was the episode too embarrassing to bring up? Did they think that by referring to it the credibility of their initial sighting would diminish? If so, the decision not to talk is a mistake as it leaves us with inevitable and still unanswered questions now that such a story has 'escaped.'

Chandler, who surely must have seen the lighthouse, never mentions it. At 4.11 am, Base Commander Colonel Ted Conrad met with Dave King, a British police officer from Woodbridge, in the forest. He was called out because Colonel Conrad decreed these things were beyond USAF jurisdiction and a "British matter." Suffolk Constabulary log books affirm the call-out. By Buran's estimate the mission was now ended. Were the men back on base or still on their four-mile chase? The British PC and USAF colonel saw only the lighthouse from the forest and seemingly identified it. Many later visits by USAF personnel to the woods must also have seen the lighthouse. Yet nobody from base is arguing the lighthouse is the UFO. That in itself is important. Police returned again to the site in daylight after a second call from Conrad. They were now shown indentations found by Burroughs. King was unimpressed yet again, noting they looked like scratch marks made by rabbits. With this, the local police lost all interest in the affair.

So how do we interpret this case? Does the story of the aborted lighthouse chase make it more – or less – likely that the Orford Ness beacon was the main culprit all along? Both Burroughs and Penniston have insisted that this theory is absurd. "I never once saw a lighthouse that flew," Burroughs told me. But unless the 1981 statements were lies, they apparently were fooled by it for some time on that same night. Yet such an embarrassing incident would make the presence of the lighthouse obvious during all subsequent investigations and sightings. The base was fully alerted to its presence. Could it ever be the culprit again?

If these men are honestly describing the close encounter phase, it is very difficult to see how this could have been caused by just the lighthouse. But why was this beacon never mentioned alongside the UFO given that both were supposedly in virtually the same point of space?

Could the lighthouse somehow have generated a bizarre visionary experience? Could the NSA experiments have created a beam of electric energy that

MEXICAN WAVE

During the summer of 1991, Mexico City was invaded by a wave of UFO sightings. Many occurred during a solar eclipse that took numerous witnesses out onto the streets of this thriving city. Amidst the darkened skies a large white light appeared, seemingly without moving. It disappeared when the sunlight returned after the eclipse.

Still photographs and several videos were taken by witnesses during the incident. The result was huge interest in UFOs, with TV specials.

Reporter Jaime Maussan collected many such images of UFOs. By far the most remarkable was a short sequence of video with a date stamped on it. This was August 6, 1997. Submitted anonymously, Maussan received it two months later.

The footage shows a large disc-shaped object in the sky in broad daylight. It is hovering above the rooftops of central Mexico City, evidently taken from the flat roof of a high-rise building. Excited voices on the audio track are heard describing the encounter. The disc then moves sideways and disappears behind a skyscraper.

The film is by quite some margin the most impressive looking, supposedly real video image of a UFO. It is so large, sharp and clear that it cannot be any form of misidentification, such as an aircraft or helicopter. The only options available are a real spacecraft or a hoax.

Is it real?

● The footage taken during the 1991 eclipse is clearly a real phenomenon. There were countless witnesses and several interlocking pieces of film

taken from different locations.

● The 1997 images are stunning and were immediately described by some UFOlogists as "the next best thing to seeing a real UFO." The craft appears to move naturally as if it is a large controlled device.

● The craft over Mexico City displays the correct perspective and even seems effected by atmospheric haze. Any hoax would have to be skilful and expensive to produce.

Is it solved?

● The eclipse sightings involve at least two objects that became visible only when the sunlight had faded. They vanished when it returned. One of them was located in the same position in the sky as the planet Venus.

196

scrambled radio reception, left the witnesses disorientated and provoked an altered state of consciousness? During that state might the lighthouse have been the source of a visionary episode that is now mixed in with all the true memories of that night?[18]

All we can say of these things is "Perhaps." What we know for sure is that the lighthouse played a far bigger role than previously imagined and seems much more likely to be somehow responsible than the alien starship legend still rife within UFOlogy and the mass media.

Physical proof?

If the events of the early hours of December 26 formed the only basis of this case, it would be an intriguing detective puzzle in need of solution. But there was a re-run less than 48 hours later which is the major reason why this case

---

This is not normally visible in daylight, causing obvious confusion.
- The second object appears to be reflecting sunlight and was in a position beyond the moon's shadow. Its appearance matches that of a high-altitude weather balloon circling the earth. Given its distance and height, this would seem stationary during the course of the few minutes it was visible.
- It seems improbable that whoever took the 1997 video would not have recognised its enormous commercial advantage. To send it to a TV station without any prospect of gaining recompense for its global profitability is difficult to reconcile. In the two months between the date on the video and its receipt, no sightings were reported to UFO groups regarding a similar

object to that on the film. Subsequent surveys failed to reveal any. Given the alleged presence above one of the world's busiest cities of a huge craft of extraordinary appearance – passing in the middle of the day – it is inconceivable that many people would not have seen it. The total absence of any such witnesses has to make this footage highly suspect. The film appears to have been taken from a building adjacent to a video production facility.

Conclusion
The UFO photography during the eclipse has adequate explanations. It involves ordinary phenomena witnessed in unusual circumstances during a unique astronomical event. The 1997 video would be proof of alien UFOs if real, but all evidence suggests it must be a hoax.

Sophisticated computer technology coupled to video photography make the creation of remarkable images such as these much less complex than might be thought. Given the publicity value of UFOs and aliens, it was inevitable that 21st century hoaxing would arrive sooner or later. The massive interest in Mexico and the personal crusade of Maussan collecting video evidence made him an obvious target for any such trickery. On the present balance of evidence, it is hard not to regard this film as a hoax.

Further reading
Analysis Report by Dr Bruce Maccabee in *MUFON Journal*, April 1998.

attracted such massive media attention. This has seen front-page newspaper headlines, TV documentaries and talk of a Hollywood movie based on the episode. Media sources rarely explore the less exciting options for what could have happened. The strength of this case lies in the assumption that it offers dramatic proof of a real alien intelligence defying NATO defences.

In fact, despite the number and quality of witnesses on the second night (December 27–8, 1980) and the physical evidence associated with this return visit, it is open to even more doubt.

Does the hard evidence left behind after night one stand up to scrutiny? The indentations found in the forest were 'secured' at first light on December 26. Conrad sent up an A-10 aircraft using infra-red heat-detection equipment. This found 'hot spots' in the forest. When Woodbridge police denounced the marks they were shown as rabbit scratches, the base commander accepted that he had now discharged his responsibility to the UK. If they were not interested, so be it. But the USAF did take evidence from the site, including plaster casts of the trace marks. This may be why some people (such as garage owner Gerry Harris and airman Geraldo) refer to scientists and men in laboratory coats wandering the forest that weekend.

These ground traces comprised three shallow holes in a rough triangle, some score marks on surrounding trees and – most significantly – a large hole punched through the tree cover above the ground. This was presumably caused either when the UFO arrived or departed – if, of course, any of these marks were really caused by it rather than being coincidental woodland features with mundane origins.

The police thought the holes were made by rabbits. So did local foresters. Vince Thurkettle from Tangham Woods showed them to me. There were so many around the forest that the chances of three forming a triangle were not fantastic. Burroughs and Penniston say the ground was frozen and so rabbit scratches were impossible. Their vehicles did not dent the rock-hard ground. But that argument is pointless since the rabbits could have made these holes days or weeks before. At best, this evidence is contentious.

The score marks on the trees are definitely solved. Colonel Charles Halt, who studied them and took samples of the sap away, said that they looked old. He was correct. The foresters told me that these notches were put there by their own axes to indicate trees next due to be felled. This part of the wood was reduced to matchwood within three weeks of the sighting – not because there was any radiation (as local rumour has long claimed), but because it was next on the list. It is also rather ridiculous to imagine that an irradiated area could be made safe by knocking down a few trees.

This leaves the hole through the tops of the trees as perhaps important. All who saw it thought it significant – a gaping swathe of branches that had vanished 20 feet off the ground. Forest worker James Brownlea came upon the area in mid-January and immediately considered the damage unusual. He reported it to his boss as possible evidence of a plane crash. Clearly, he thought it out of the ordinary.

The plane crash idea was reasonably rejected for the sort of reasons cited by Penniston. The woods were too dense not to leave massive evidence of the inevitable destruction. But next morning, on Brownlea's return to work, this part of the woods was already being felled and the hole had disappeared. Brownlea said that the area *was* due for cutting, but he felt the speed of felling after his discovery was strange. The Forestry Commission admit it happened, but it was coincidence. Yet if the hole in the trees attracted the attention of a man who worked in Rendlesham every day, then its status has to be thought important.

James Easton points out that the testimony of Burroughs and Penniston suggests that the UFO did not depart skyward at the point where the traces were found. But as I read their stories it might have 'landed' here – if, of course, any physical object was ever present to land. Or is this hole in the trees mere happenstance? Did its discovery somewhere near the spot where the UFO appeared serve as a hook to fix in the minds of the witnesses the point where the object had been? After all, locating any one part of an unfamiliar forest in the dark would not be easy. Finding odd marks may just have led the witnesses astray.

Again, all we can say is that without proof of the strangeness of this phys-ical evidence little can be deduced. Burroughs took his own plaster casts of the holes in the ground. Charles Halt has one, too. To date, they have provided no analysis reports to say if these traces have been properly studied, e.g. to compare with rabbit scratches. Photographs of the site were taken by the USAF later that weekend. Halt says these came out "fogged." We are left balancing possibilities. The hard evidence can only be termed inconclusive.

Star witness

Colonel Charles Halt is the highest rank USAF witness to this case. In 1980, he was Deputy Base Commander. With a degree in chemistry and a masters in business administration, Halt was a career airman. He joined the USAF in 1964, served in Vietnam and Japan, was promoted to full Colonel and Base Commander at Bentwaters soon after the UFO sightings, and later went on to run bases in Korea and Belgium.

Halt was charged with responsibility for nuclear weapons, so was a highly respected officer whose involvement in this case clearly had no impact on how he was perceived by the USAF hierarchy. That he was a witness was known from early days. He collated all the statements and wrote a memo to the MoD on January 13, 1981, on the recommendation of Squadron Leader Donald Moreland. Moreland served as a liaison between the American staff and the MoD, and asked Halt to file a report as part of this process.

Despite being promised by the USAF and the MoD that this January 13 case summary would never be released, an administrative error in the USAF saw the amazing one-page account sent out to the American UFO group CAUS in June 1983. It had been denied both to Brenda Butler and myself for two years, during which time our pleas for information from Whitehall met with replies that simply evaded all questions.

In my situation, direct requests about Rendlesham were met with case reports on other MoD incidents that were possibly sent to keep me happy and put an end to my digging into Rendlesham. When I did not give up, I finally received (April 13, 1983) a statement from the MoD clearly based on the then unreleased Halt memo confirming that strange lights were seen in the forest. These were "unexplained," I was told. But the report itself was kept secret until its submission to CAUS two months later.

This MoD case summary has been the source of much confusion because Halt appears wrongly to date the two separate events. This hampered early study. The witnesses largely agree on the correct dates and their 1981 forms concur, but these USAF witnesses remained silent for more than a decade. Several have still not gone public. It seems that Halt went from memory, not their then recent statements, when compiling this famous memo two weeks after the incidents. In 1997, he suggested to me that he used the report as a teaser for the MoD, assuming a full enquiry would inevitably follow such a serious matter.

On release of the Halt memo, the USAF told CAUS their copy had been destroyed and thanked the MoD for giving them one to release! If true, then we have the absurd situation that British citizens were denied access to a report despite it covering events in the UK and being filed by our own MoD. Yet

UFOlogists in the USA had no trouble obtaining the document via Whitehall. When I appeared at the MoD in London in August 1983 carrying a copy of the newly released memo, I asked if I was breaching the Official Secrets Act by my intention to publish. Its veracity was confirmed, Brenda, Dot and I were cleared to release it, but the MoD categorically denied sending a copy to the USAF for release. So far as Whitehall were concerned, they had stood by their promise to Colonel Halt never to make this document public.

The Halt memo summarises events of December 26. It has no reference to the close encounter phase now claimed by Burroughs and Penniston. It describes the physical traces, but not the existence of photographs, samples or plaster casts. Halt told me that he would have made these available had the MoD followed up his report. But, to his astonishment, it was simply ignored by the UK. Nobody contacted him at all. It seems the MoD, like the Woodbridge police, filed this account and just forgot about it!

Halt's memo also makes no reference to some further astonishing evidence secured during the second night's encounters. This was a live tape of the events as they unfolded. It was recorded in the forest by Halt on a portable dictaphone and includes eighteen minutes of discussion made on site assessing the physical traces left after the first night. This was suddenly interrupted by a return visit of the UFOs. A series of USAF personnel – Halt included – describe the unfolding encounter 'live' in hushed tones.

In August 1984, a copy of this taped record was released by a former officer at Bentwaters who had retained a copy when he moved to command a base in Texas. Why he did so remains a puzzle. The release of this stunning tape and MoD memo firmly established Halt as the key player in the second night's events, but the colonel declined entreaties to give public interviews until he retired from the USAF in 1992. His story comes from interviews given after then to our *Strange But True?* investigation (1994), *OMNI* science magazine (1996), in conversation with me (1997) and an Internet chat room question-and-answer session (1998).

The return

Colonel Halt was aware of the events from the early hours of December 26. On arrival for duty at breakfast that Friday, the base was already alive with talk about the UFO. He knew the men involved in the chase for the lights, considered them reliable, but was no fan of UFOs. However, he expected an investigation and was surprised that nothing had been put in the 'blotter,' the base log of events for the night. He was told that people were "embarrassed" and nobody knew what to say. Halt ensured that this was immediately put right. The idea to use the innocuous term "unidentified lights" in the records was his.

On the Saturday evening (December 27), there was an officers' base party at Woody's Bar. This was the main event on Halt's mind as that weekend unwound. But it was not so for some of the men on base. According to Burroughs, he could not shake a presentiment that the UFO would return. He spent off-duty time in the woods. Several airmen who were not witnesses add that the rumours had spread so quickly that quite a few would-be UFO observers went out to Rendlesham over the next few nights on what we might term impromptu 'skywatches.' A few minor sightings of lights were reported by this route.

Off base things were happening, too. At least as many civilian witnesses exist for the events of late on December 27 as do for the sightings 48 hours before. From witness statements and report forms collated by Brenda Butler and me – mostly between 1981 and 1984 – here are some examples.

Tony Sorrell, a plant manager at Thetford, describes how "For some unknown reason I looked up into the air above my house ... I saw this triangle-shaped object moving along. It was like lightly frosted glass. As it moved, the stars above it dimmed ... It made no noise and was very sharply defined." The object headed off southwards towards Rendlesham Forest.[19]

It is worth noting that this statement from Tony Sorrell is signed before Jim Penniston gave his description of the object. Nor could Penniston have seen it before he spoke. Yet the account of a frosted glass triangle is remarkably similar to Penniston's opaque smoky glass object. It is also interesting that Sorrell's claim that something just made him look up is supported by several other witnesses that night. It matches well the 'feeling' in the air that Burroughs had.

Rather than dismiss this as coincidence or some psychic phenomenon of undefined origin, another possibility occurs. Some people are sensitive to the onset of a thunderstorm. What they are doing is detecting the ionisation in the atmosphere caused by the charged particles associated with the seething electricity. If a natural phenomenon was producing a glowing light with an ionisation field surrounding it, this, too, may have given its presence away without witnesses knowing what they were 'sensing.'

Gordon Levett lived at Sudbourne, on the coast north of Orford Ness and on the very edge of the forest. As the house was isolated, he always put his dog into a large kennel in the garden to serve as a guard against burglary. Levett was doing so late this night when, as he puts it, "my attention was aroused by some unknown means." The dog was similarly alerted and both stared northwards at an object that was silently crossing the sky at low height. Levett likens what he saw to an overturned mushroom with an eerie phosphorous glow. A greenish tinge was behind the white. His sketch is actually more akin to a shallow triangle or cone, and he insists that the object paused briefly directly above his head at a height of maybe 30–50 feet. Then it continued towards Woodbridge base and was lost above the trees of Rendlesham.

Next day, the dog was showing signs of great distress. Its condition deteriorated – and within a week or so it was dead. The vet suggested poisoning, but no source was ever found.[20]

From within Rendlesham Forest, a courting couple saw the glowing mass descend into trees. Dave Roberts and his girlfriend were situated off the Orford to Woodbridge road when they saw this happen. There was no sound and they got out of the forest quickly. Then the couple saw military jeeps rush to the site.

All of these events seem to coincide with the first return visit that Saturday night reported by airmen from the USAF base. At around 10 pm on December 27, a security patrol out near Woodbridge East Gate had a virtual re-run of the first night's events. This close similarity between the two sightings and the haphazard bit by bit way in which data emerged across many years added to the confusion over this case. Parts of the two sightings were commonly mixed and it was often hard years later to know which night a particular story related to.

We know two members of this four-man security patrol on 27 December because they have spoken in a limited way. Airman Greg Battrom gave an interview to US satellite station CNN in 1985. With him on the patrol was his supervisor, Sergeant Adrian Bustinza, who spoke to American church minister and UFOlogist Ray Boeche and colleague Scott Colborne in 1984, albeit reluctantly. Ray sent me a copy of these phone interviews a few days later. Bustinza was out of the USAF, but in government service and clearly not keen on going public with his tale. However, his story matches the one later given by Battrom very well.

Battrom told how the patrol saw white lights over the forest that were not like any known aircraft. They got permission for CSC to go into the woods and saw "a fire" in a clearing. This is the same location as where Burroughs and Penniston saw the Boast house "glowing" on December 26 and again suggests a connection with the lighthouse, which is directly in line of sight. However, once again there was a close encounter phase that makes little sense of it being just a beacon light.

Battrom told how as he and his team walked towards the glow the hair on his neck and arms began to stand up. A loud noise described as a "thrumming" was also emitted by the object. The glow had a ground mist surrounding it. This pulsed with different colours. There was no definitive shape. The nearer to this phenomenon the witnesses got, the stronger the electrostatic field. Their radio communications also began to suffer interference. Battrom had endured enough by this point and was glad to quit the forest.

Bustinza tells how they were led towards the clearing by a white light over the trees. Once here, they saw a ghostly scene that included mist or fog rolling around. Perhaps this was simply localised mist as the weather data for Suffolk that Saturday suggests pockets of mist were possible and the low-level clearing seems a good gathering place.

Whatever the case, Bustinza says the fog was "thicker at the centre than at the edge – a yellow mist about two to three feet off the ground. It was like dew, but yellow … (inside) there was a red light on top and several blue lights on the bottom, but there was also an effect maybe like a prism – with rainbow lights scattered about. It was weird." He recalls how they decided to get out of there fast and knocked down a fence between the clearing and the forest in their retreat. I saw and photographed that fence in 1983. It instantly came to mind when Bustinza told his story the following year.

Bustinza added that at closest approach he was terrified. As the thing moved through the trees in silent majesty, a curtain of intense static charge rippled through the forest – "We were wearing regulation headgear, but I could see one man's hair standing up on his head just like it was a wire brush." At that moment, the forest burst to life with deer and rabbits fleeing like crazy. "In a funny way this rooted me back in normality. I was glad of it," he reported.

The investigation

Battrom, Bustinza and the other two men regrouped on the edge of the forest. The on-duty Flight Chief, Lieutenant Bruce Englund, arrived (possibly in a jeep witnessed by Dave Roberts). It seems that quite a few other people who heard the radio chatter unofficially made their way to the forest. They spread out over the next few hours, and some may have had later sightings at various points throughout the woods.

One man who went of his own volition and did have a second encounter was John Burroughs. His precognition was being vindicated by events that he was not about to miss.

Englund could see nothing from the edge of the forest. No strange lights were visible. But he could tell that these men were frightened, so decided to do the unthinkable and interrupt the officers' party at Woody's. At 10.30 pm he arrived, pulled Base Commander Ted Conrad and deputy Charles Halt aside and explained that the UFOs were back. Halt said (1996) that the lieutenant was "as white as a sheet."

Conrad was due to give a speech, so asked Halt to take charge. The colonel should find out what was really causing these men to see UFOs. Neither officer believed such things were real. Halt gathered a small team, including Englund. He noted (1996): "Rumours about what Penniston and Burroughs had seen were beginning to circulate out of control. I was determined to put an end to this UFO nonsense."

Halt was certainly ready. He stopped off to change into a utility suit, picked up an office tape recorder because, as he told me (1997), the night was quite windy and he felt note-taking would be difficult. He also selected an airman who had been on base several years, was a nature lover and knew the forest well. This, Halt assured me, (1997) was one reason he is confident the light-house was not a factor.

In addition to these men, there was Sergeant Munroe Nevilles of the base Disaster Preparedness team. He served as photographer and knew how to work a Geiger counter. This expediency was thought necessary because of the infra-red emissions from the forest detected by the previous day's A-10 flight.

Englund had gone back to the woods to help set up 'light-alls,' portable searchlights. It was decided that they would make very sure this time that they could see whatever was out there. It might be that the "thrumming" reported by Battrom from his position across the forest was one of these if he is mis-remembering when it was heard, for on the Halt tape the throbbing of the gener-ator is audible in certain parts.

Burroughs (1994) recalls that when he reached the end of the logging track where the light-alls were taken, "there was already a lot of activity going on." He began a search for the UFOs. Soon enough he would find them, only this time "the lights were different. They were blue glows flying around in the air. And they were also beaming light rays down towards the ground."

By the time Halt and his team – now fully equipped – met with Englund by the clearing, there were already problems. The three light-alls were working intermittently. Halt told me they were considered reliable, but proved unusually bothersome that night. Even the replacement they sent out for failed.

Throughout the four hours they were in the woods, they never had more than one out of three light-alls working at any time – and this never ran for more than a few minutes. Halt is sure this was most unusual.

Halt's first tape message after arrival records: "Ah, 150 feet from the initial – er, I should say, suspected impact point. Having a little difficulty. We can't get the light-alls to work. Seems to be some kind of mechanical problem." He notes they were sending for another and in the meantime "We're gonna take some readings with the Geiger counter and, ah, chase around the area a little bit …"

ALIEN AUTOPSY

In 1993, rumours circulated around British UFOlogy that film of an autopsy on one of the aliens from the 1947 Roswell crash had surfaced. British rock video entrepreneur Ray Santilli had discovered it in the USA. Philip Mantle of BUFORA (British UFO Research Association) maintained a rapport with Santilli, who eventually purchased footage. One of Santilli's former colleagues, rock-singer-turned-UFO-enthusiast Reg Presley, revealed its existence on a TV interview in early 1995.

At the same time, Mantle was shown dimly-lit images allegedly filmed in a tent near Roswell depicting an alien body. Other film supposedly included scenes of a site visit by President Truman.

In May 1995, Santilli held a private screening in London. Invited UFOlogists and media guests came from all over the world. Better quality black and white film was now available from stock Santilli said he bought from a retired US photographer who had held on to these images for 48 years, but needed to sell to help his family.

Deals were quickly done by Santilli's company with the world's media, and a commercial video release set. The poor quality 'tent footage' was excluded, although it appeared on a Milton Keynes independent video called *Penetrating the Web 2*, a sequel, but with different participants, to *Penetrating the Web 1*. That was made in Milton Keynes in 1993 with BUFORA backing and featured interviews with UFOlogists such as Jenny Randles.

Mantle and Santilli agreed an exclusive premiere of the new footage at the August 1995 BUFORA conference in Sheffield. No samples were made available for BUFORA to research in advance. Plans for an open debate on the evidence (already being challenged by many leading UFOlogists) were disallowed by BUFORA.

At the behest of Santilli, amazing security precautions were taken on the audience before, during and after the Sheffield screening. The new footage showed scientists dissecting organs from a human-like, bald-headed figure in a white room. Other images showed metal bars with symbols on what purported to be Roswell wreckage.

Unfortunately, the aliens were not the small, egg-head beings claimed to have been seen by alleged eye witnesses. Moreover, the wreckage was not that described by people in Roswell, such as Jesse Marcel, Junior.

The film generated huge publicity world-wide with several TV specials, but was disowned by the vast majority of the UFO movement, many writing articles arguing why it must be a hoax. The cameraman has never gone public to answer questions, but Santilli stands by his own innocence.

Radiation

No aspect of this case is more controversial than the radiation recorded at the site. The taped record features much talk by Halt, Nevilles, Englund and others as they prod around the 'secured' trace area from the first night. They appear to struggle with the Geiger counter, whilst statements like "This thing is about to freak" pop up amidst debates about the scale in use. Readings are taken from the 'pod marks' (the indentations), the tree score marks and, as the night progresses, all over the fields in a much wider area.

At the same time, the men are using a 'starlight scope,' an infra-red binoc-

---

BUFORA was heavily criticised for associating itself with untested evidence and assisting its promotion.

Is it real?
- The Roswell case did occur. Some stories claim bodies were recovered.
- Film stock (from a leader tape) provided by Santilli was tested and could date to 1947.
- No obvious major anachronisms have been found within the room seen on the autopsy footage.

Is it solved?
- The leader tape was blank and had no images to confirm absolutely that the footage dates to 1947.
- Witness stories about the form of the aliens are not consistent with the images on the film, but these accounts are the only evidence that alien bodies were recovered at Roswell. If they are discredited by images on the film, there is no

support for any bodies at Roswell to be autopsied.
- In January 1999, members of a video production team came forward to explain how they faked the tent footage. A model head and a child were used, a barn near Milton Keynes serving as the 'tent.' The farmer interrupted filming and was adopted to play President Truman dressed in clothes from a scarecrow! The makers of *Penetrating the Web 2* may not have known its origin.
- Santilli admits he sent early unprocessed film supplied by the cameraman for it to be 'cleaned' by this Milton Keynes team, but they failed to find any retrievable images, so faked their own. He later withdrew the tent footage from marketing and never sent further stock – including that which produced better images – to this company. The new film may thus be real, Santilli argues.

Conclusion
This case began with a dubious pedigree and has steadily declined. The main players such as Santilli may well be sincere and reporting only what they know, and BUFORA is innocent of all but possible misjudgement.

Motives of the unknown cameraman cannot be defined in his absence. The footage has always been suspect to UFOlogists because of its inconsistency with the Roswell case. To date, little hard evidence has been provided by independent analysis that comes close to establishing it as authentic film of a 1947 autopsy.

Santilli has issued riders for use on video sales that say he believes it real, but cannot prove it to be so. The admission that the first images to appear (the tent footage) are a hoax must make anyone doubt the sense of regarding this case as proof of aliens in the absence of a lot more evidence.

ular that shows up residual heat sources optically. Several glows coming from trees – most strong immediately near the landing site – are described. It seems hard to interpret any real strangeness to this. Trees do retain heat.

Most of the time, the Geiger counter gives "just minor clicks" – as they are described – suggesting normal background count. But slightly higher readings occur in the centre of the traces and a point where the ground displays a 'blast effect.' Halt records on tape: "We've found a small blast – what looks like a blasted or scuffed up area here. We are getting very positive readings."

These readings display the unfamiliarity of the team with the equipment. Numbers like "seven-tenths" and "half a millirem" are quoted, but with no helpful time base. It is a bit like saying that a car's speed is 50 miles without adding "per hour" as a time base. There is discussion on correct use of the device.

It is worth noting that nobody seems to be taking serious precautions against possible irradiation. They take samples, photographs, make sketches and stay out there for hours. But only at one point does Halt suggest they "put the gloves on." This implies that radiation is not believed by the men to be a threat.

Halt also describes the gap in the trees "15 to 20 feet up," where something has punched a hole through the pine cover. He adds that some branches an inch or so across that appear "freshly broken" are on the ground underneath. This again confirms the potential significance of that evidence.

As for the starlight scope, this frequently gives glowing images and its use is widely discussed. At one point Halt reports: "We get a high radioact ... er, a high reading. You're sure there's a positive after effect?" "Yes, there is definitely," one officer confirms. "... We are getting an indication of a heat source coming out of that centre spot, which will show up on the scope."

"Heat or some form of energy," Halt cuts in. "It's hardly heat at this stage of the game."

The way that Halt clearly backs off midstream from using the word 'radioactive' is interesting. But there is no doubt the team believe they are detecting some sort of anomaly. Halt publishes details of some of these findings in his report to the MoD on January 13. He records "Beta/Gamma readings of 0.1 milliroentgens ... with peak readings in the three depressions and near the centre of the triangle."

But is this at all strange?

I doubt it. In 1983 when I first saw the Halt memo, I checked with plant biologist Dr Michele Clare. She reminded me that radiation can accumulate in a forest and especially in holes where pine needles drop off. At best interpretation, the figures described by Halt are only twice normal background and not a real anomaly.

Interviewed by *Strange but True?* in 1994, physicist Alan Bond was slightly more impressed, although not overwhelmed. He thought the figures might indicate levels a little greater than those left in the UK after the 1986 Chernobyl nuclear accident in the USSR. This had led to health checks on food.

However, Nick Pope, former head of the MoD UFO department, says he took the data to experts consulted by the ministry in radiation matters as part of an official enquiry. They asked where on earth he had got these results as they indicated a level ten times normal. In such a situation, the fact that Halt and his team suffered no ill effects is notable – and even more so the apparent lack of physical illness to Burroughs, Penniston and Cabansag since radiation decays exponentially and they were supposedly on site within moments of its deposition.

This implies that the 'ten times normal' claim is surely flawed. Science writer Ian Ridapth added to the evidence that this is so. He went to some trouble tracking down Giles Cowling at the Defence Radiological Protection Service, whom Nick Pope consulted. He then checked with the Radiological Protection Board and the manufacturers of the Geiger counter used by Halt. All indicated that the equipment was not suitable for background readings in a forest, and that the figures cited are even so at the lower end of the scale. As Dr Clare long ago told me, nothing more than a couple of times normal levels are involved – and even here these levels are open to doubt due to the mechanics of the device itself.

James Easton further points out that the readings in the field beyond Boast's house taken later in the night are not massively less than those found in the indentations. Unless all of this forest were seriously irradiated, this implies that the readings at the trace site were not very significant.

It is safe to conclude that no radiation levels of any proven importance were recorded.

Eye in the sky

All of this probing of the "impact site," as Halt calls it, ended at 1.48 am. As Halt records on the tape, events really now began to move in the forest. By then, his team had been out there a couple of hours and seen nothing odd. Halt told me (1997) that he definitely saw both the Orford Ness and Shipwash lightship during this time and he certainly should have done from this clearing. Ian Ridpath suggests that there was patchy mist. Did that make the lighthouse invisible from the site for the first few hours? Was the lighthouse shining through gathering mist the source of the initial 'misty UFO' sightings made by Battrom, Bustinza and others? If so, what are we to make of their claims about an electrostatic field?

Halt flatly refutes this theory. He told me there was no significant mist at all that night. It was clear and cold, with many stars visible. In fact, he explained, he had the tape recorder with him because of the quite notable wind, and this would have been inconsistent with any gathering mist. This all needs to be taken into account as the strange events begin to unravel.

As Halt reports into the tape: "Zero one forty eight (1.48 am) – we're hearing very strange sounds out of a farmer's barn-yard animals – very, very active. Making an awful lot of noise."

Had one of the pockets of men in the forest, such as Burroughs, alerted the Boasts' dogs by straying too close to the 'farmhouse'? Certainly, many voices start to talk at once and someone has evidently seen a light out towards the east. Halt cuts in on tape: "You just saw a light? Where? Slow down. Where?"

Then seeing it, the airman reports: "Right on this position here. Straight ahead in between the trees. There it is again! Straight ahead of my flashlight beam. There it is!"

This sequence is very important as it shows that the light was intermittent – pulsing in, then out, then becoming visible again. The airman was pointing straight at the Boast house. And straight at Orford Ness Lighthouse, which does, of course, pulse in and out as described. Had the mist cleared just enough to make the beacon suddenly visible? Had its dramatic appearance caught these men unawares?

Halt confirmed he saw this new light. A conversation recorded on tape shows that it was reddish and "small." Through mist, the lighthouse could indeed take on an orange colour.

Colonel Halt has offered more detail as to what his team saw at this point, flatly rejecting the lighthouse theory (1994) – "A glowing red sphere. It looked like the sun when it first comes up in the morning, although it had a black centre and it pulsated as though it were an eye winking at you."

In 1996, he added that he could not see a specific shape, but it then did something strange. It created the weird effect also reported on the first night and by Bustinza's patrol three hours before. It seemed to make the Boast house glow as if on fire – "I did not know what to think. I was quite concerned. In fact, we were all quite concerned." Minutes later, when they passed the house in pursuit of the light, they realised this 'house on fire' impression was an illusion caused by the glow of the light beyond the property. Halt considered waking the occupants, but "I thought it prudent to keep our presence low," so they went on by.

In his 1994 interview, Halt likened this strange effect to "molten metal dripping off." He added that it appeared to come near and then recede, just as I have seen the lighthouse do from this spot many times when it brightens and dims during its cycle of rotation.

This recurrent 'house on fire' image may be a key to the entire case. I certainly never witnessed anything like it from the site. Nor have the many other people who visited the spot since that weekend. So why did the house appear to glow? That it was admitted by Halt to be an illusion is important. He knew the source was well beyond the house shining on it, possibly reflecting off its windows. The lighthouse was beyond the house in line of sight. I never saw it shine off windows or appear nearly as bright as these stories allege. But it was there and "winking like an eye" as the beacon rotated.

Another clue is that Halt and the team were using the 'starlight scope' to look for infra-red radiation, evidently pointing it hither and thither, as the tape records. This magnifies the image and shows heat sources as a glow. The lighthouse was, of course, a heat source as was the Boast house. Viewing these through the scope might have enhanced, if not created, the illusion being described.

There are other vital pointers. On the tape, Halt gives the bearing of the object as "110 degrees," says it seems "out towards the coast" and remarks on its pulsation. Indeed, from the one point on the tape where the light (now with a "yellow tinge in it") appears, then vanishes and reappears, some comment by the men makes it possible to time the 'periodicity' of the UFO and compare with the lighthouse. In every detail, the 'winking eye' matches the lighthouse – direction, distance and frequency of pulsation.

The way in which the object appeared suddenly, seemed like a molten red orb, then became more yellow and also "more clearly defined" all fits with a beacon light gradually emerging from ground mist. Its distortion and coloration would become less prominent. The local mist may well be clearing to refine the image. The distance between the men and the light also closed as Halt ordered his team to leave the site where the traces were found and head off in pursuit of the UFO. Either way, the distortion to the lighthouse would have become less apparent as time progressed just as the UFO seems to do at this point of the case.

It might also be worth recalling the description from earlier that night when Bustinza likened the thing to a yellowish glow seen through mist and throwing off lights like a prism. Mist can create a mirage and there are differences in refraction of light through air with varying temperatures. Lights can change colour when viewed through mist and the path of the light rays be bent to create a showering or prismatic effect. I have seen this when taking part in a skywatch at Weir, in a UFO-rich 'hot spot' of Lancashire. Sitting in a car by a reservoir, an eerie-looking light appeared on top of the water. Only after some minutes was its origin clear. A bright star had just risen and was being refracted by a small pocket of mist on the water surface. In the dark I had not noticed the mist: it was only obvious when the star rose above it. When it did, the star's origin became self-evident as it was no longer being distorted and discoloured.

From this experience, I can imagine what a bright lighthouse beacon might look like through mist. You can also get some idea by the way a pencil 'bends' when put in water: light is refracted differently through air and water, causing the illusion. You can even see it on hot days when an illusion of a 'pool' appears on the road ahead, just like the fabled oasis that travellers see in a desert. What you and they are viewing is no hallucination. It is part of the sky 'dislocated' because of this 'refraction' effect caused by the temperature difference in the air close to ground level.

This evidence does strongly support the view that the lighthouse – plus mist – may be an answer. You may only waver in accepting the theory when you hear Halt on the tape describe in awed tones how, as they walked towards the glow, "It appears to be moving a little bit this way. It's brighter than it has been. It is definitely coming this way … pieces of it are shooting off … There's no doubt about it – this is weird."

Are these chilling words a description of a fascinating prism illusion caused by the lighthouse through ground mist or something more bizarre? Of course, remember Halt said categorically in 1997 that he saw both Orford Ness and Shipwash lightship as well as the UFO that night. If so, this answer fails.

Is there any cause for doubt given Halt's obvious insistence on this matter? Unfortunately, there is. Halt took part in an Internet question-and-answer session. James Easton sent me this extract.

Might he have been deceived by the Orford Ness lighthouse? Halt was asked. In reply, Halt noted: "First, the lighthouse was visible the whole time. It was readily apparent, and it was 30 to 40 degrees off to our right. If you were standing in the forest where we stood at the supposed landing site or whatever you want to call it, you could see the farmer's house directly in front of us. The lighthouse was 30 to 35 degrees off to the right (south) and the object was close to the farmer's house and moving from there to the left through the trees."

This is vital evidence for two different reasons. Here, Halt is not describing the Orford Ness lighthouse as he seems to believe. These directions from the site locate the Shipwash lightship to the south. Is this again telling evidence that the UFO *was* the lighthouse? Did the airmen believe they were seeing the light-house when in truth they were viewing the lightship? That seems to be the implication. But what of Halt's claim that he saw both? Moreover, if the distant Shipwash lightship was visible to Halt, this does not support the presence of local mist distorting the appearance of the lighthouse – then far closer – unless, of course, the mist was only found in one direction and not to the south where the lightship was located.

Further mysteries

The main UFO disappeared this way, according to Halt (1994): "There was what was like an explosion. Not with any loud noise. Just an explosion of light, and it disintegrated and broke up into three to five different objects. Then it was gone." Making the lighthouse fit this, even via a mirage, is tough.

You can also add the testimony of Michael Simms, who lived in residential quarters on Bentwaters as son of an officer. Looking with friends towards Woodbridge, he saw the following: "It was an enormous object ... mesmerising. Then it flew away from us, picking up speed ... We tried to chase it towards the forest ... Then it split into three parts, each light flying off in different directions and disappearing."

Meanwhile, Halt and his team had set off across the clearing, past the Boast house, through fields towards Capel Green and got wet wading through a 'creek.' It was now after 3 am.

In my 25 years' experience, long duration sightings of this kind have a rational explanation just about all of the time. Real UFOs are a very transient phenomenon. They do not stay in the sky for hours. This is an important reason why I believe that this second night will turn out to have an IFO solution.

On tape, Halt records what the men could now see. The big light had disappeared, even though he claims – but never mentions on tape – that the lighthouse remained. "Now we have multiple sightings of up to five lights with similar shaped orbits. Seemingly steady now, rather than pulsating or glowing with red flashes." But these soon changed to doing a "grid search – sharp, angular movements, pulling very high Gs (as tight turns on aircraft are termed)."

He elaborated (1994) that "somebody noticed these objects in the sky. When we looked to the north about 10 degrees off the horizon there were two large objects – looked almost like the moon, only not as large. They were round, and as we watched they changed shape from round to elliptical ... It was almost like an eclipse." It is not clear if this was all seen visually or via binoculars and the starlight scope, either of which could have distorted point sources of light in this way. Many sightings of strange UFO shapes in the sky turn out to be bright stars seen distorted via some optical lens, such as binoculars or view-finders.[21]

In 1996, Halt describes the process thus: "The objects appeared elliptical and then they all turned full round ... They were stationary for a while and then they started to move at high speed in sharp angular patterns." By all accounts, these lights stayed in the sky for hours, more or less in the same place.

I find it difficult to believe from my experience with similar cases that these were anything other than bright stars. Several candidates in that part of sky have been identified, such as Vega. The "angular movements" are the biggest clue. UFO researchers are familiar with an effect called "autokinesis." A person staring at a static bright light against a dark sky for some minutes will commonly report that it starts to move in this jerky way. It does not move, but the eyes do, making the image on the retina zip about in this "grid search" manner. Our brains are used to interpreting the movement of images on the retina as either motion by ourselves or by an object. If the brain knows you are standing still, it will decode these patterns as angular motions by the object you are watching. It is a psychological phenomenon very well understood and a feature of human perception. I suspect that these lights remained stationary and the motion was created by autokinesis. If that interpretation is correct, the lights were surely just stars.[22]

Halt patiently rejects the theory and has support from Woodbridge garage owner Gerry Harris. He had just returned with his wife from a Saturday night out and saw moving lights in the forest.

Harris reports: "I stood watching by the gate. Three separate lights moved up and down. I thought, 'If they are aeroplanes, they are going to crash into the trees.' All of a sudden, one of them went down behind trees. Then after a few minutes it went up again like a bat out of hell." As he is not sure of the exact time, this may be a description of the Bustinza sighting earlier that night. But Harris was not using binoculars or optical aids. He adds that following the disappearance of the lights, the base burst into life, saying, "I could hear it from here – the lorries and the people shouting and vehicles starting up. At that time of night it was unusual." This may better fit the time immediately prior to the arrival of Halt in the forest.[23]

More definite as a civilian back-up to Halt's sighting is the story of Sarah Richardson. She places the events between 1 and 3 am, and notes how from her house at the time (in Woodbridge) she had a fine view towards the base and forest. She was familiar with aircraft, helicopters, searchlight beams and all the multiple activities of a busy USAF base. That night was different and very memorable – "Three bands of light appeared over the woods to the side of the runway. They were star-like, and they were bright, coloured red, blue and yellow. One was in the north, two were in the south."

Sarah confirms there was no mist. The night was sharp, frosty and clear. She was looking at Sirius, a star pointed out to her by a friend who knew astronomy. These lights were positively not stars. She thought for a time they were fireworks connected with a Christmas party. But they remained far too long.

After watching for some minutes, Sarah opened her window and leant out. The glow was "intense" and "kind of unreal." She contemplated the possibility of helicopters, but had seen many of these and they looked very different. Besides, there was no sound and their position, as later confirmed in daylight, meant they would be hovering dangerously low amidst dense forest. To her, the oddest feature was the recurrent colour changes. Stars, of course, scintillate and seem to switch colour. Sarah believes they remained in the sky for well over an hour, possibly two, until around 4 am. Then they just "shot straight off."[24]

Laser beams

Halt and his team were also seeing an object to the south. On the tape we hear the following: "Zero Three Fifteen (03.15 am) – now we've got an object about ten degrees directly south … but the ones to the north are moving away from us."

A voice cuts in (believed to be Sergeant Munroe Nevilles), "It's moving out fast."

As the sighting develops, Halt's voice begins to show real strain. It is obvious from the tape that this is the point where all thought that these lights were somehow explicable flew out of the window. "Mm – they're both heading north ... Hey! ... Here he comes from the south. He's coming towards us now ... Now we observe what appears to be a beam coming down towards the ground – *this is unreal!*"

Halt explained (1994) why this was the defining moment. "I wondered if this was a friendly probe or a weapon. Were they searching for something? At this stage, my scepticism had definitely disappeared. I was in awe." In 1996, he added: "At one point, it appeared to come towards us at very high speed. It stopped overhead and sent down a small pencil-like beam, sort of like a laser beam. It was an interesting beam in that it stayed the same size all the way down. It illuminated the ground about ten feet from us and we just stood there in awe wondering whether it was a signal, a warning or what it was."

The beam "just clicked off" and the light "moved back towards Bentwaters and continued to send down beams of light, at one point near the WSA" (weapons storage area). Halt knew this because despite unusual static interference that dogged radio reception all night (causing them frequently to switch channels) he heard excited conversations from the base. People there were seeing light beams raining down on the WSA – the place where, unknown to the slumbering, Christmas-partying British people, nuclear weapons were stored. Halt estimates maybe 30 on base saw these lights.

None has come forward, but one man traced over the Internet in 1998 by James Easton was non-commissioned sergeant Randy Smith, of the ATC (Air Training Command). He was then on duty in the base tower. Smith recalls the furore that night: Major Bob Ball of security even asked for permission to cross the active runway at one point, something undertaken only when an emergency situation is occurring.

Located by the nuclear bunkers, Smith saw the commotion as people watched the skies – "I asked what everyone was looking at and they pointed out three objects that appeared like stars to the naked eye. Binoculars were being passed around, and when I had my turn I saw very clear images of three triangular-shaped craft that were hovering a few miles away and above treetop level." Meanwhile, the radio was alive with talk about the light-alls, strange UFOs and beams. Note here how 'stars' became 'triangles' when viewed through binoculars – precisely the optical distortion problem noted before.[25]

John Burroughs was not with Halt during the encounter with the laser beams, but he saw them from his position near where the mobile lights had been left behind. (1994) "All of a sudden, a blue light streaked out of the sky and went past us. It shot through the open window of our truck ... Then all the light-alls went out. There was no noise. It just went by us in a blur." This begs the thought, 'Did Halt see a beam coming down to the ground or one going up into the sky?' He had left the light-alls far behind. These searchlights were still operating in his absence. What if someone had pointed them skyward?

Halt does not support this idea. Burroughs told me (1989) that there were beams like this all over the forest. He heard men say they passed through trees and the metal of a jeep, emerging the other side as if defying physics. But they had a real substance. When one passed within inches of Burroughs, he felt the cold rush of air that filled the gap after the beam vanished. It was like a slip-stream.

There are only a couple of brief comments on the tape after the lasers. Halt led his team back to base sometime around 4 am. The colonel says his men were cold, wet and very tired having been up half the night traipsing across fields. As the men returned home, the star-like lights in the north were still present, further indicating to me that indeed this is precisely what they *were* – stars. The final proof comes, I believe, when Halt tells us that back on the base, at around 6 am dawn began to paint the horizon in the east. "The objects (in the north) were still in the sky. However, it was getting light and they were getting faint."

Had these things been helicopters (or indeed UFOs), one would imagine daylight would make their structure more obvious. But stars – being simply light – do fade as the sky brightens, eventually getting swallowed up. Given this comment, it is hard to regard these things in the northern sky that were visible for several hours as anything but stars. I think the balance of evidence in this regard is overwhelming.

Conclusions

What are we to make of this fantastic case? It has taken almost twenty years, but we have gradually worked our way towards a solution. There are some things we can say with reasonable certainty and others more open to debate. However, the prospect that this case is one of the biggest UFO encounters on record is, I think, much harder to support in the light of these developments. The idea that alien craft appeared in the forest is even less sustainable. Rendlesham Forest as a case may not be dead, but it does need a doctor.

That this case has an extraterrestrial origin was what attracted UFOlogy and the media to it and will no doubt continue to do so. Two new books are due about it. But the fact that this interpretation was fed to the UFO community from the start is disturbing. Some of this was doubtless because of rumours spreading around the base in its wake. With hundreds of men and dozens of witnesses involved, it was inevitable that the truth would be distorted by constant retelling.

Moreover, various excited airmen fascinated by UFOs (as many people are) could not resist going out into the forest sky watching. The air of expectation

---

UP IN THE AIR

On April 28, 1999, European media reported that a mid-air encounter had occurred between a UFO and a Debonair BAe 146 aircraft flying a party of SAAB businessmen from Sweden to Humberside Airport in the UK.

The incident (as later checks revealed) was at 28,000 feet on February 3, 1999, (5.25 pm GMT) and occurred 58 miles off the Danish coast. It involved (Press stories alleged) a UFO "as big as a battleship" that was a long cylinder with "rows of square portholes."

On landing, the crew reported this to the MoD, who were not investigating. Media stories continued that British military radar had picked up a target five minutes after the sighting, implying this was vindication of the UFO. As the latest in a number of mid-air encounters this seemed a serious matter, but UFOlogists soon discovered the story that the public were never told.

Is it real?

● The four-man crew were highly skilled individuals who filed a 'mandatory occurrence report' and clearly believed their sighting was possibly a threat to their jet. They acted with propriety.

● Over 50 very similar cases have been traced by UFOlogists involving British aircraft alone since 1991. Several of these were tracked on radar. Many involved dark cylindrical objects.

● Three other aircraft supported the claim that a UFO was seen that night off the coast of Denmark.

Is it solved?

● The official report filed with the MoD was supplied to David Clarke on April 30. It contains no reference to the object being "as big as a battleship," a phrase possibly borrowed from a 1998 Press report about another sighting over the North Sea that UFO investigators showed to have no substance. Nor is there mention of "portholes." The crew merely describe "an area under the aircraft illuminated by an incandescent light." This persisted for a few seconds.

● No military radar detected the UFO. Indeed, no radar in the UK could possibly pick up an object so far from Britain, as the Civil Aviation Authority confirmed. The target was actually seen on the aircraft's own weather radar, which is configured to detect extremely large masses such as mountain ranges and storms. It could not possibly detect a cylindrical UFO.

---

and credibility must have ensured some things were misperceived. Of course, there remains the story of Jim Penniston and how he touched a landed craft with symbols etched onto it. If this memory is a true reflection of a real event that part of the case defies explanation, but it is complicated by the distortions, suggesting an altered state of consciousness during this period. Can we take what the witnesses say about that time literally? Even then, Penniston's hypnosis story tells us there were no aliens present in the woods, only humans from the future!

If alien visitors are hard to square with the evidence, what really happened? There is no question that some major elements of the case can essentially be blamed upon mundane phenomena. The Orford Ness lighthouse does seem to have been witnessed on both nights, despite claims to the contrary. I find it hard

● Swedish UFO researcher Clas Svahn worked on the matter on May 15, 1999, with the Head of Military Radar at the Swedish Department of Defence. CD Rom records of all radar covering the location (south-west of Jylland and near Ramme) were scoured. Nothing was on there except the aircraft. Taking the statements of the other crew into account (all Scandinavian airline flights), it is believed the UFO was miles high in the sky and beyond radar reach, certainly not close to the BAe 146. Defence experts propose it was a fireball meteor creating a luminous trail as it burned up.

Conclusion
This is a classic case of the Press jumping to conclusions from half the facts. Once the full facts are available, a seemingly baffling UFO

turns into a solvable event.

There are several almost identical precedents dating back to a July 1948 encounter involving an Eastern Airlines DC-3 over Alabama. Here, a bolide meteor emitting a plasma trail triggered a sighting that shook the Pentagon.

Jenny Randles probed an incident that created TV reports and newspaper headlines in 1996 when civil aviation sources rather hastily suggested a UFO might have been seen. This involved a British Airways' Boeing 737 on a flight from Milan to Manchester.

BA 5061 was passing over Whaley Bridge, Derbyshire, at 6.48 pm on January 6, 1995, making its final descent just above thick cloud when the two-man crew saw a wedge-shaped glow (with either portholes or lights on its edge). It passed close by the aircraft. No passengers saw it, and there

was no wake turbulence. Investigation revealed that radar only tracked the BA jet.

Again, this object was almost certainly a bolide miles high and posing no threat to the 737. Many witnesses have described the trail of glowing debris of such a fireball as resembling a craft with portholes. Distance is frequently misjudged to be much closer than in reality. But media sensationalism ensured the truth did not emerge. The same may well occur with the Debonair encounter. Such is how UFO legends are born.

Further reading
Something in the Air.
Randles, J. Robert Hale, London, 1998.

not to believe the lights in the northern sky seen by Halt were probably stars. The very first object that fell into the forest and seen by Steffans on December 26 appears to match the meteor seen at 2.50 am. To his credit, astronomer Ian Ridpath saw some of this long before any UFOlogist did.

However, can we now dismiss the entire case in these terms? Therein lies the real problem.

What were the coloured lights seen from the forest edge, where neither lighthouse nor lightship were visible? If Halt and the others truly did see both the lighthouse and the lightship during the sighting, what was the main light flying about the woods?

Could the lights that projected 'laser beams' have been high-flying military craft belonging to the RAF? There was a British exercise off the coast that weekend. Bentwaters was alerted to it. There may have been traffic nearby. Halt reports twice contacting RAF Watton to see if they had radar returns over the forest. Their logs record one of these calls at 03.25 on December 28. Could this contact with the RAF have resulted in aircraft from the exercise being routed over the woods to take a look, though unbeknown to Halt at the time? Of course, the descriptions offered do not resemble helicopters using searchlights. It is hard to imagine one of these hovering over Halt in the midst of a quiet wood without its presence being obvious. Helicopters are far from quiet.

There seems a good chance that some of the close encounters witnessed that weekend were a result of the lighthouse distorted through patchy ground mist, resulting in a prismatic mirage. This might create an illusion of a fiery object breaking apart. The witnesses were looking right at the lighthouse during much of this affair and it is difficult to believe it was not a vital factor. Had the tapes or reports at the time indicated that any of the witnesses were seeing the light-house *as well as* the UFO, things would have been different. But at no point do they do that. However, can we accept that this is all that happened? This light-house was in the forest night after night before and after, and it is improbable that everyone out there was unaware of it. Local civilians certainly were and, of course, they saw strange things that weekend too.

But what could make the lighthouse take on a dramatically new appearance? Was it excitement, expectation and disorientation? Did the presence of mist create a distortion of this light to the point that it became totally unrecognis-able? Or was something occurring on Orford Ness that played a part?

The experiments were real. By the testimony of witnesses unconnected with this case they did from time to time result in electrical disturbance and sightings of glowing lights in and around the Ness. It has to be possible that such a test occurred during that Christmas weekend, and when witnesses saw UFOs, as a consequence they were seized upon by the powers-that-be to hide the truth. Any glowing lights may have combined with the lighthouse to sow seeds of confusion. Those in the know would hardly have chosen to clarify this confusion. And no government source would bother to investigate Halt's report.

Perhaps the electrical fields associated with the experiment contributed to the witness deception. Note that many people in the area spoke of 'sensing' something as if they were aware of a charge in the atmosphere. Unless there is a huge amount of deception involved, we have to take seriously the stories of radio interference and that the electric powered light-alls were not working properly. Then there are the claims of Bustinza, Battrom, Burroughs and Penniston about the intense electrical fields that swept through the forest, causing hair to stand on end and skins to tingle. Did this electric charge also disturb the local wildlife as reported on at least three occasions in the forest? These stories have a consistent common denominator – an electric field – the very thing connected to the radar experiments in this same area.

If there was such a strong electrical field, could that have contributed to the distortions of perception experienced by these witnesses? This is no wild speculation. There has been extensive research since 1977 by Canadian brain specialist, Dr Michael Persinger, investigating the effects of strong electromagnetic radiation on the human mind. He has demonstrated conclusively in laboratory work that they can create both physiological effects (tingling sensations, headaches and nausea) and psychological changes that constitute an altered state of consciousness.

From this work, Persinger has even concluded that what he calls 'transients' – naturally occurring floating clouds of energy – can create ionisation in the atmosphere and be perceived as glowing UFOs. If a sighting of one of these is accompanied by an altered state, the witness may 'visualise' a close encounter. Persinger believes that alien abductions, for example, might occur when a witness misinterprets a meeting with a mundane light as an alien spaceship, and imagination and cultural expectation take over, transforming what is seen through a visionary process.[26]

Of course, if there are transients – and these can occur naturally – perhaps they can result from experiments such as those on Orford Ness. Perhaps the presence of the lighthouse acting as a trigger could, in the presence of a strong energy field, provoke a close encounter vision, especially if that light is assumed to be a UFO. Could this explain nagging doubts about the close encounter phase of the encounter?

Whilst some puzzles remain, we can probably say that no unearthly craft were seen in Rendlesham Forest. We can also argue with confidence that the main focus of the events was a series of misperceptions of everyday things encountered in less than everyday circumstances. This is not to challenge the integrity of witnesses such as Halt, who offer credible testominy. But one question outstanding is whether the misperceptions were entirely a result of psychology and sociology or were side effects stimulated by some nefarious experiment.

NOTES AND REFERENCES

Chapter One
1. "Once upon a time in Aurora." Jeff Gorvetzian. *Fortean Times* 115 (October, 1998), p. 36.
2. "Phantom Aerial Flaps and Waves." Watson, Nigel. (*Magonia Magazine* special publication, 1987), pp. 3–5.
3. *The Unidentified*. Clark, Jerome and Coleman, Loren. Warner Books, (New York), 1975, p. 133.
4. "The Airship Hysteria of 1896–97." Robert Bartholemew in *The UFO Invasion*. Prometheus Books (New York), 1997, p. 17.
5. *Dallas Morning News* (Texas), April 16, 1897.
6. *UFOs and Alien Contact*. Bartholemew, Robert and Howard, George. Prometheus Books (New York), pp. 24–25.
7. Ibid, pp. 25–29.
8. *San Francisco Examiner*, December 6, 1896.
9. *Springfield News* (Illinois), April 15, 1897.
10. *The UFO Encyclopedia*. Clark, Jerome. Omnigraphics (Detroit), Vol. 1, p. 59.
11. *The Books of Charles Fort*. Fort, Charles. Dover Books (New York), 1974, p. 470.
12. *Stockton Evening Mail* (California), November 27, 1896.
13. Ibid.
14. "The Airship Hysteria of 1896–97." Bartholemew, p. 21.
15. "Airship steals calf." *Kansas City Times*, April 27, 1897, p. 1.
16. *Flying Saucers – Serious Business*. Edwards, Frank. Lyle Stuart (New York), 1966.
17. *Anatomy of a Phenomenon*. Vallee, Jacques. Henry Regnery Company (Chicago), 1965.
18. *UFOs: Operation Trojan Horse*. Keel, John. Putnam (New York), 1970, pp. 76–77.
19. "Hamilton's Airship Hoax – Kansas 1897." Bob Rickard. *Fortean Times* 20 (February, 1977), pp. 5–8.
20. *Atchison County Mail* (Missouri), May 7, 1897, quoted in *The Airship File*. Bullard, Eddie. (Unpublished manuscript, 1982), p. 104.
21. *The UFO Encyclopedia*. Spencer, John. Headline Books (London), 1991, p. 137.
22. *Hamilton's Airship Hoax*. Rickard, p. 7.
23. *The UFO Encyclopedia*. Clark, Vol. 1, p. 62.
24. *Once upon a time in Aurora*, p. 38.
25. *Dallas Morning News* (Texas), April 19, 1897, quoted in *The Airship File*, pp. 226–28.
26. *The UFO Encyclopedia*. Clark, Vol. 1, p. 262.
27. "Aurora Spaceman – RIP?" Eileen Buckle, *Flying Saucer Review* Vol. 19, 4 (July/August, 1973), pp. 7–9.
28. *The UFO Encyclopedia*. Clark, Vol. 1, p. 262.
29. Ibid.
30. *Aurora Spaceman – RIP?* p. 8.
31. Ibid.
32. Ibid.
33. *Once upon a time in Aurora*, p. 38.
34. *New York Times*, February 27, 1979, quoted in *The UFO Encyclopedia*, Clark, Vol. 1, p. 263.
35. Quoted in *Once upon a time in Aurora*, p. 38.
36. *Aurora Spaceman – RIP?* p. 9.
37. *UFOs and Alien Contact*, pp. 204–5.

Chapter Two
1. *Cosmic Crashes*. Redfern, Nicholas. Simon & Schuster (London), 1999.
2. *UFOs and Alien Contact*. Bartholemew, Robert, and Howard, George. Prometheus Books (New York), 1998, pp. 205–6.
3. See "Ghost Fliers." David Clarke, *Peak and Pennine Magazine*, May 1999, pp. 24–27.
4. "Moors Plane Crash Riddle." Paul Whitehouse and David Clarke, *Sheffield Star*, March 25, 1997, p. 1.

5. A copy of South Yorkshire Police's Major Incident Log of the incident was obtained by Sheffield UFOlogist Martin Jeffrey from a contact within days of the incident.

6. Police Log entry no. 32, 10.54 pm.

7. Live Internet question-and-answer session featuring Max Burns on Visitations, June 6, 1998.

8. Interview with Marie-France Tattersfield, March 25, 1997.

9. Interview with Barbara Ellison, March 25, 1997.

10. Interview with Detective Inspector Christine Wallace, Hammerton Road Police Station, Sheffield, March 27, 1997.

11. Interview with Mrs Dronfield, March 29, 1997.

12. Interview with John Littlewood, February 22, 1999.

13. Interview with Chief Inspector (now Superintendent) Christine Burbeary, BBC1 *Mysteries*, November 1997.

14. See *The Howden Moor Report*, Clarke, David, and Jeffrey, Martin (IUN 1999), pp. 29–33.

15. Police Log, 1.51 pm, March 25, 1997.

16. See "The Night of the Phoenix." *UFO Magazine (UK)*, May/June 1997, pp. 8–11, 58–9.

17. "The Sheffield Incident: A Flying Triangle Incident." Burns, Max. PUFORI Internet Website, 1998, p. 1.

18. *The Sheffield Incident*, p. 4. Flight Lieutenant Philip Inman of RAF Linton-upon-Ouse wrote in a letter of March 1998 that "from records it is confirmed that Royal Air Force Linton-upon-Ouse was closed at the time of the reported/suspected incident. The search and rescue operation did not include any aircraft or personnel from this Unit or the satellite stations."

19. *The Sheffield Incident*, p. 18.

20. Letter from Nick Pope, May 19, 1999.

21. Information provided by Mike France, Peak District Mountain Rescue Organisation, 1998.

22. Ibid.

23. "Mystery of the fire-crash ghost plane." Harry Cooke, *Daily Express* (London), March 26, 1997, p. 26.

24. "Seismic Alert: Sonic Event, South Yorkshire." March 24, 1997, 22:06 UTC. Press Release from the British Geological Survey, Edinburgh, March 27, 1997.

25. See *The Isle of Lewis Mystery*. Chapter 7.

26. *The Sheffield Incident*. pp. 2–3.

27. "Crash that Never Was." David Clarke, *Sheffield Star*, March 24, 1998, p. 6.

28. Information provided by Derbyshire Police, 1998.

29. *South Yorkshire Police Log 2* (Supplementary), entry number 32, logged 11.39 am, March 25, 1997. This witness has made no further statement regarding this sighting, and has failed to respond to inquiries by several investigators.

30. *UFO Revelation*. Matthews, Tim. Blandford (London), 1999, p. 109.

31. *Hansard*. Written Questions, March 23, 1998.

32. Ministry of Defence, Written Answers, March 25, 1998.

33. Ministry of Defence, Written Answers, April 7, 1998.

34. Interview with Helen Jackson, MP. *Origin Unknown*, Granada TV, November 1998.

35. Interview with Flight Lieutenant Tom Rounds, October 27, 1998.

36. Interview with RAF PRO Alan Patterson, May 1998.

Chapter Three

1. *Look North West*, BBC Television, June 1991.

2. All early data from the BUFORA case notes compiled by Stanway, R., 1973. Available in BUFORA archives. (16 South Way, Burgess Hill, Sussex, RH15 9ST.) Web site www.bufora.org.uk

3. OS references from the site surveys by SCUFORI, 1984, and personal visits, 1988. SCUFORI case report available via the BUFORA archives.

4. Ibid BUFORA, 1973.

5. *UFOs: A British Viewpoint*. Randles, J. and Warrington, P. Hale, London, 1979.

6. *Fire in the Sky*. Randles, J. BUFORA, August 1989.

7. *UFO Reality*, pp. 203–6. Randles, J. Hale, London, 1983.

8. Latest information on Ball Lightning via TORRO, PO Box 164, Richmond, Surrey, TW10 7RR. Web site: http://www.zetnet.co.uk/oigs/torro

9. Letter to Moffatt, M., SCUFORI, dated April 3, 1984, from Murone, V.,

Chief of the Reports
Branch, Directorate of
Aerospace Safety, Norton
Air Force Base, California,
US.

10. Press features include *The
Times*, London, January 12,
1973, and *Bletchley
Gazette*, January 19, 1973.

11. *Kilroy*, BBC1, November
1987. Follow-up interviews
by Randles, J. Nov
1987–Feb 1988.

12. Live comment on *Kilroy*,
op. cit.

13. This was included in the
original version of his book
Open Skies, Closed Minds.
Simon & Schuster, London,
1996. When challenged on
it by Jenny Randles
(December 1995), Pope did
not recognise the case and
was unsure where his
details came from.

14. *UFO Reality*, 1983, op. cit.,
pp. 201–3.

15. Telephone interview,
February 1988.

16. On March 13, 1991, Campbell sent a draft of the
article to Peter Day, who
forwarded to me. Day
declined permission to
include photos in the
article, and wrote on March
19, affirming this to the
editor. I also wrote to the
editor requiring correction
of their statements about
BUFORA. Neither of us got
a reply. The article appeared
as "Fireball by Day –
Steuart Campbell explains
another 'UFO' film," in the
*British Journal of Photog-
raphy*, April 4, 1991. Day's
copyright photo was
included. Campbell later
claimed that Day's rejection
was irrelevant as "copyright
law" gave permission to use
the photo for "criticism and
review."

17. Articles about the crash are
in *Aircraft Illustrated*, June
and July 1992.

Chapter Four

1. *UFOs and Alien Contact*.
Bartholemew, Robert, and
Howard, George.
Prometheus Books (New
York), 1998, p. 63.

2. Quoted in *The Mothman
Prophecies*. Keel, John.
Saturday Review Press
(New York), 1975, p. 39.

3. "North America 1966:
Development of a Great
Wave." John Keel, *Flying
Saucer Review*, Vol. 13, 2
(March/April, 1967),
pp. 3–9.

4. The flap of 1988 is
described in detail in *Fly-
by-Night*, by David Clarke,
published as an IUN Investigation Report, 3 volumes,
1988.

5. "UFO Hijack Terror." Chris
Anderson, *Daily Star*
(London), January 22,
1988, p. 1.

6. Investigation by Ernie Still
for BUFORA, quoted in
The August Report by
David Clarke, published by
the IUN in 1987.

7. *The Hull Report*. Gary
Anthony, IUN Investigation
Report, 1987. "Mystery sky
sight remains a mystery."
Susan Critchley, *Hull Daily
Mail*, December 10, 1987,
p. 1.

8. "Probe launched into UFO
claims." David Welford,
Derby Evening Telegraph,
August 18, 1987, p. 3.

9. *The August Report*, ibid.
*Phantoms of the Sky:
UFOs, a Modern Myth?*
Clarke, David and Roberts,
Andy. Robert Hale
(London), 1990, pp. 30–31.

10. Letter from C. R. Neville,
MoD Sec (AS) 2a, to David
Clarke, March 28, 1988.

11. Letter from Nottingham
Weather Centre to David
Clarke, October 13, 1987.

12. "'Ello, 'ello ... that's a
UFO." Peter Wilson,
London *Evening Standard*,
January 5, 1988.

13. Personal communication
from Mike Wootten,
January 1988.

14. "May the force be with
you." *Daily Mirror*
(London), February 8,
1988, p. 3.

15. "Police saw UFO hover
overhead." Keith James,
Sheffield Star, February 6,
1988.

16. *Sheffield Star*, February 8,
1988. *Fly by Night*, ibid.

17. Ibid.

18. *UFO Brigantia*, May/June
1988, p. 7.

19. Letter from Flight Lieut C.
J. Hamilton, RAF
Finningley, to David
Clarke, March 1, 1988.

20. Letter from G. C. Dennison,
Airport Director, Leeds
Bradford Airport, to David
Clarke, March 14, 1988.
Letter from D. Coulson,
Senior Air Traffic Control
Officer, East Midlands
Airport, March 15, 1988.

21. See *UFO Brigantia*,
March/April 1988, p. 28
and May/June 1988, p. 6.

22. See pp. 138–9.

23. Letter from the MoD to
David Clarke, January 12,
1989.

24. "Eee by gum lad, UFOs
over Barnsley." David
Clarke, *UFO Brigantia*,
July/August, 1988, pp. 5–8.

25. *UFO Brigantia*,

March/April 1988, p. 29.

26. *Phantoms of the Sky*. Clarke and Roberts, pp. 34–5.

27. *Fly-by-Night*, ibid, Vol. 3

28. Letters from RAF bases at Linton-upon-Ouse, Leeming, Binbrook and Church Fenton, April 1988. See *Fly-by-Night*, ibid.

29. *The May 1988 Stafford UFO Flap*. Clive Potter. Report to BUFORA/IUN, 1988.

30. Ibid.

31. *Fly-by-Night*, ibid, Vol. 2.

32. *UFO Brigantia*, September/October, 1988, p. 8.

33. *UFO Revelation: The secret technology revealed?* Matthews, Tim, Blandford (London), 1999, pp. 143–51.

34. *Phantoms of the Sky*. Clarke and Roberts, p. 37.

35. *UFO Revelation*, Matthews, pp. 159–68.

36. See, for example, *The Flying Triangle Mystery*. Omar Fowler, Phenomena Research Association (Derby), 1996.

37. "Has BAe's Halo slipped?" Tim Matthews, *UFO Magazine (UK)*, January/February 1997, pp. 50–51.

38. "The truth is up there." David Clarke, *Sheffield Star*, February 28, 1998, p. 3.

39. *Encounter with the Silent Vulcan at Uttoxeter*. Clive Potter. Report to BUFORA and the IUN, 1989. *Phantoms of the Sky*. Clarke and Roberts, p. 38.

40. *Flight International*, November 19, 1988.

41. *UFO Revelation*, Matthews, pp. 194–200.

41. Ibid., pp. 98–100; *Sunday Telegraph*, December 18, 1994; *Air Forces Monthly*, November 1994.

42. *Cosmic Crashes*. Redfern, Neck. Simon & Schuster (London), 1999, pp. 149–52.

43. *UFOs: A Federal Case*. William Spaulding, in *The Age of the UFO*, Orbis Publications (London), 1984, pp. 34–39.

44. A memo from CIA director Walter Smith in 1952 mentions a study which concluded "that the problems associated with UFOs appear to have implications for psychological warfare as well as for intelligence and operations." It suggested a board meeting of CIA experts to discuss the possible offensive or defensive utilisation of these phenomena for psychological warfare purposes. See *Phantoms of the Sky*. Clarke and Roberts, pp. 109–133.

45. "Social and Cultural Factors influencing belief about UFOs." Dr Phyllis Fox, in *UFOs and the Behavioural Scientist*, ed. Haines, Richard. Scarecrow Press (New Jersey), 1978.

Chapter Five

1. *Daily Mirror*, August 23, 1983.

2. *Cracoe UFO Report*, YUFOS, 1986, p. 10.

3. Op. cit., p. 11.

4. *Retrievals*. Four Winds Productions, 1996.

5. *Cracoe UFO report*, YUFOS, 1986, p. 8.

6. *Cracoe, The Evidence*. Mark Birdsall, 1986, p. 20.

7. Op. cit., p. 21.

8. *Cracoe UFO Report*, p. 14.

9. Op. cit., p. 32.

10. Op. cit., p. 38.

11. *Cracoe, The Evidence*. Mark Birdsall, 1986, pp. 66–67.

12. Op. cit., p. 52.

13. *YUFOS Journal*, Vol. 2, No.5, Nov 1983, p. 5.

14. *Cracoe, The Evidence*, Vol. 2, Mark Birdsall, 1986, p. 126.

15. *Craven Herald*, August 26, 1983.

16. *Craven Herald*, September 2, 1983.

17. *Quest*, Vol. 4, No. 4, Sept–Oct 1985, pp. 7–11.

18. *UFO Brigantia*, No. 23, Nov–Dec 1986, p. 20.

19. *Quest*, Vol. 6, No. 4, Sept–Oct 1986, p. 13.

20. *UFO Brigantia*, No. 23, Nov–Dec 1986, p. 21.

21. *Quest*, Vol. 6, No. 4, Sept–Oct 1986, p. 7.

22. *UFO Brigantia*, No. 24, Jan–Feb 1987, p. 7.

23. Op. cit., p. 7.

24. Op. cit., p. 7.

25. *Yorkshire Evening Post*, January 5, 1987.

26. *Yorkshire Evening Post*, January 5, 1987.

27. *Alien Investigator*. Dodd, Tony, Headline 1999, p. 6.

28. Op. cit.

29. *Phantoms of The Sky*. Clarke, D and Roberts A, Hale 1990, p. 12.

30. *Unopened Files*, Quest Publications, No. 3, 1997.

Chapter Six

1. Stornoway Coastguard Incident Action Report, Incident Number 0190, October 26–27, 1996, provided by the HM Coastguard Press Office.

2. Interview with Norman MacDonald, BBC1 *Mysteries*, November 1998.

3. Interview with Fred Simmons, BBC1 *Mysteries*, November 1998.

4. Coastguard Log, p. 4.

5. Interview with Senior

Coastguard Watch Officer Ian Lindsey, December 1998 and March 1999.

6. "The Lewis Island Incident: A Connection?" *UFO Magazine (UK)*, January/February 1997, p. 21. "Big bang over Scottish Island." Jenny Randles, *Northern UFO News*, 177 (October, 1997), pp. 15–16.

7. *Cosmic Crashes: The incredible story of the UFOs that fell to Earth*. Redfern, Nick. Simon & Schuster, London, 1999, p. 168.

8. Interview with Ian Lindsey, ibid.

9. "US Boffins probe riddle of Ness sky-blast." Taylor Edgar, *Stornoway Gazette*, (Isle of Lewis), November 7, 1996, p. 1.

10. "'Top Secret' riddle over sky-blast." Taylor Edgar, *Stornoway Gazette*, October 31, 1996, p. 1.

11. Ibid.

12. Ibid.

13. Interview with Chris Murray on BBC1 *Mysteries*, November 1998.

14. "Minch hunt for Mystery Plane." *The Press and Journal* (Aberdeen), February 28, 1961; BUFORA Investigation report ref. no. 61008.

15. *The Scotsman* (Edinburgh), November 4, 1997, p. 7.

16. "UFO Riddle No. 2." *Stornoway Gazette*, October 31, 1996.

17. Interview with Ian Lindsey, ibid.

18. *Glasgow Herald*, November 11, 1996, p. 2; *The Scotsman*, November 4, 1996.

19. *The Scotsman*, November 4, 1996.

20. "Mid-air explosion, Isle of Lewis." *Hansard* (Written Answers), October 14, 1997.

21. "The 5,000 mph spy." *Daily Mail* (London), December 15, 1992, p. 15.

22. "Secret spy jet blows its cover." John Ingham, *Daily Express* (London), December 15, 1992.

23. *UFO Revelation: The Secret Technology Exposed?* Matthews, Tim. Blandford, London, 1999. pp. 219–20.

24. Interview with Paul Beaver, *Jane's Defence Weekly*, BBC1 *Mysteries*, November 1998.

25. Interview with Major Roy Denton, of the British Army, Benbecula Testing Range, Outer Hebrides, March 1999.

26. Interview with Ian Lindsey, ibid.

27. Interview with Dr Jacqueline Mitton, March 1997.

28. Interview with Dr Mark Bailey, March 1999.

29. *Cosmic Crashes*. Redfern, pp. 165–66.

30. Interview with Dr Mark Bailey, March 1999.

31. Private e-mail from Dr Mark Bailey to Dr David Clarke, March 30, 1999.

32. Interview with Ian Lindsey, ibid.

33. See Jenny Randles, chapter 9, pp. 178.

34. "Blinding trail from a shower of meteors." *Daily Mail* (London), June 12, 1998; "Lights in the sky leave Britain in grip of Z-files." *Daily Express* (London), July 12, 1998.

35. Information from Glenn Ford, British Geological Survey, Edinburgh, March 1999.

36. UK Earthquake Monitoring 1996/97 Eighth Annual Report, British Geological Survey (Edinburgh, 1997), p. 12; "Seismic Alert: Sonic Event, NE Scotland, 23 September 1997, 0758 UTC." BGS Press Release, September 23, 1997.

37. "Moorlands mystery over fall from sky." David Clarke, *Sheffield Star*, December 9, 1997, p. 2.

38. Nick Redfern reproduces a letter from Andrew Bates of Secretariat (Air Staff) 1a at the MoD dated November 18, 1996, referring to the case. He said the MoD's involvement "only concerned the deployment during the search phase of military SAR assets and we have no evidence to support any of the media theories about the cause of the incident." See *Cosmic Crashes*, Redfern, p. 167.

39. "Aliens ruled out as mystery distress calls alert satellite." *Press and Journal* (Aberdeen), November 18, 1996, p. 2. "Mystery helicopter found in Atlantic." Iain Maciver, Atlantic Press Agency release, December 17, 1998.

40. Interview with Norman MacDonald, BBC1 *Mysteries*, November 1998.

41. *UFOs and Alien Contact*. Bartholemew, Robert and Howard, George. Prometheus Books (New York), pp. 205–6.

42. Ibid.

43. *Open Skies, Closed Minds*. Pope, Nick. Simon & Schuster, London, 1997.

44. Letter from Nick Pope to Dr David Clarke, May 19, 1999.

45. Letter from Nick Redfern to Andy Roberts, February 10, 1999.
46. *Cosmic Crashes*. Redfern, pp. 162–63.
47. *Alien Investigator: The case files of Britain's leading UFO detective*. Dodd, Tony. Headline Books, London, 1999, pp. 227–28.
48. Private correspondence with Jen Topping, *Stornoway Gazette*, February 3, 1999.

Chapter Seven

1. "Silent Saucers over Mosbro." *Derbyshire Times* (Chesterfield, Derbyshire), June 22, 1962, p. 22.
2. "Do you think these things are flying saucers?" *Sheffield Telegraph*, June 20, 1962.
3. Ibid.
4. Interview with Stuart Dixon, Mosborough, April 6, 1999.
5. "Flying Saucer with an orange glow filmed by Sheffield man." *Sheffield Star*, August 20, 1962.
6. "This is what I saw." Keith Graves, *Sheffield Star*, August 29, 1962.
7. BUFORA file 62009, Alex Birch case.
8. Report by John Adams in BUFORA case file, dated March 10, 1963.
9. Report by Alan Watts in BUFORA case file, dated September 21, 1962.
10. "Flying Saucers: The evidence runs on straight lines." Waveney Girvan in *The Star Weekend Magazine* (Sheffield), September 1, 1962.
11. PRO Air 2/16918: "Correspondence with Mr A. Birch about UFOs at Mosborough, near Sheffield." Letter from A. Birch (senior) dated July 2, 1962.
12. PRO file, letter from R. H. White to A. Birch, July 23, 1962.
13. Internal S6 memo written by R. H. White, February 11, 1963.
14. Interview with Alex Birch, Nottinghamshire, October 22, 1998.
15. Ibid.
16. Internal S6 memo from Flight Lieutenant A. Bardsley, DDI (Tech), September 24, 1962.
17. Letter to A. Birch (senior) from R. H. White, S6, September 25, 1962.
18. Memo from Flight Lieutenant Bardsley to S6, September 25, 1962.
19. File 7824 Project Blue Book, National Archives, Washington, DC. Conclusion reads: "Insufficient data for evaluation. Negatives not with prints. No request made for photo analysis."
20. Interview with Alex Birch, Nottinghamshire, November 8, 1999.
21. See "Hoaxer confesses after ten years." *Flying Saucer Review*, Vol. 18, 6 (November–December 1972), p. 2.
22. "Sheffield man confesses … I hoaxed the world with UFO picture." Malcolm Tattersall and Heather Smith, *Sheffield Telegraph*, October 6, 1972.
23. Stuart Dixon interview, April 6, 1999.
24. Birch interview, November 6, 1998.
25. *Sheffield Telegraph*, October 6, 1972.
26. "Lift off for UWO." David Clarke, *Sheffield Star*, February 9, 1999.
27. "No kidding this time … My flying saucer picture is genuine." Eileen Brooks, *Yorkshire Post* (Leeds), March 5, 1999.
28. Interview with David Brownlow, December 3, 1998.
29. Interview with Stuart Dixon, December 4, 1998.
30. Interview with Stuart Dixon, April 6, 1999.
31. Letter from A. Birch (senior) to BUFORA, April 25, 1963.
32. Interviews with Alex Birch, October 22 and November 6, 1998.
33. Letter from Alex Birch to Nina Pendred, editor of *Alien Encounters*, March 13, 1998.
34. See *Star Children*. Randles, Jenny. Robert Hale, London. (Frequent references under the heading 'psychic toys'.)
35. See "The Mothman Prophecies," Keel, John, *Saturday Review Press* (New York), 1975, reprinted 1991, for numerous references to a sound like a woman screaming, reported by UFO and poltergeist witnesses in the USA during the wave of 1966–67. It also features in the Rendlesham Forest case (see p. 185)
36. Interview with Alex Birch, November 6, 1998.
37. Ibid.
38. *Electric UFOs: Fireballs, electromagnetics and abnormal states*. Budden, Albert. Blandford (London), 1998.
39. *UFOs and Alien Contact*. Bartholemew, Robert and

Howard, George. Prometheus Books (New York), 1998, pp. 251–68.

40. Ibid., pp. 267–68.

41. Quote from the Birch website, www.ufo-images.ndirect-co.uk 1999.

42. Ibid.

43. See "Are those fairies real after all?" June Ducas. *Sunday Telegraph* (London), July 12, 1998.

Chapter Eight

1. "Phantom Helicopters Over Britain." Clarke, D. and Watson, N. FUFOR, 1985.

2. *UFO Retrievals*. p. 120. Randles, J. Blandford, 1995.

3. There are numerous sources for the basic myth. Readers are directed to the following: *A Covert Agenda*. pp. 111–124. Redfern, N. Simon & Schuster, 1997; *Earthlights*. pp. 228–231. Devereux, 1980; *UFO Reality*. pp. 152–153. Randles, J. Hale, 1983; "Another One Bites The Dust." pp. 18–20, Clarke, D. *UFO Brigantia*, May 1989; "Britain's Roswell." pp. 12–13. Randles, J. *Sightings*, Vol. 1, No. 3, 1996; "The Night The Mountain Exploded." pp. 9–12, Randles, J. *IUR*, Winter 1996; "UFO Crash In North Wales." pp. 34–37, *UFO Magazine*, Sept/Oct 1996; *Alien Investigator*. pp. 205–210. Dodd, T. Headline, 1999.

4. *Daily Telegraph*, January 24 and January 25, 1974; *The Times*, January 25, 1974; *The Guardian*, January 25, 1974.

5. *Border Counties Advertiser*. January 30, 1974.

6. *Wrexham Leader*. January 25, 1974.

7. *UFO Retrievals*. pp. 180–181. Randles, J. Blandford, 1995.

8. *Places of Power*. pp. 114–116. Devereux, P. Cassell, 1990.

9. Personal communication from Jenny Randles, April 8, 1999.

10. *A Covert Agenda*. pp. 111–124. Redfern, N. Simon & Schuster, 1997.

11. "UFO Crash In North Wales." pp. 34–37. *UFO Magazine*, Sept/Oct 1996.

12. *A Covert Agenda*. p. 123, Redfern, N. Simon & Schuster, 1997.

13. *Phantom Helicopters Over Britain*. Clarke, D. and Watson, N. FUFOR, 1985.

14. Personal communication from Nick Redfern, October 28, 1998.

15. Letter from University of Leicester Department of Astronomy, March 25, 1974.

16. Details from British Geological Survey, Edinburgh.

17. Interview with Pat Evans, January 31, 1998.

18. Interview with Pat Evans, January 31, 1998.

19. BGS Field Diaries, ref Mrs Pat Evans C75, 1974.

20. Interview with Huw Thomas, January 31, 1998.

21. BGS Field Diaries, ref C64.

22. Interview with Huw Thomas, January 31, 1998.

23. Gwynedd Constabulary Major Incident Log, January 23, 1974.

24. RAF Valley Mountain Rescue Team Log, January 25, 1974.

25. Personal communication from Dr Ron Madison,

December 23, 1997.

26. MoD Letter to R. Foxhall, July 14, 1998.

27. *Cosmic Crashes*, pp. 124–144, Redfern, N. Simon & Schuster, 1999.

Chapter Nine

1. Letter to Mike Wootten, BUFORA. August 3, 1985.

2. BUFORA case files and various local newspapers. December 27, 1980.

3. British Astronomical Association report by Spalding, G. and Mason, Dr J., 1981.

4. *Pennine UFO Mystery*. Randles, J. Grafton, London, 1983.

5. *UFO Crash Landing?* pp. 35–8. Randles, J. Blandford, London, 1998.

6. *Mysteries of the Mersey Valley*. Randles, J. and Hough, P. Sigma Press, Cheshire, 1993.

7. "Blythburgh and Sizewell UFOs." Johnson, F. *Flying Saucer Review*. Vol. 21, No. 5, 1976.

8. *The Complete Book of UFOs*. pp. 110–19. Hough, P. and Randles, J. Piatkus, London, 1997.

9. The Laxfield case emerged by chance as this chapter was being compiled, but is supported by many other sightings.

10. *A Secret Property*. Noyes, R. Quartet, London, 1985.

11. *The Melbourne Incident*. Haines, Dr R. Palo Alto, USA, 1987.

12. Vince Thurkettle is quoted in *The Times*, London. October 3, 1983. "A Flashlight in the Forest." Ridpath, I. *The Guardian*. Manchester, January 5, 1985.

13. Interview by Brenda Butler

and Dot Street, February 1981.

14. Interview by Jenny Randles, October 1983.

15. Interview for *Strange But True?* May 1994.

16. *Sky Crash*. Butler, B., Randles, J. and Street, D. Grafton, London, 1986.

17. *Complete Book of Aliens and Abductions*. Randles, J. Piatkus, London, 1999.

18. *Electric UFOs*. Budden, A. Blandford, London, 1998.

19. Witness statement to BUFORA, November 1983.

20. Interview by Jenny Randles, October 1983. Also signed legal affidavit by witness.

21. A similar effect is noted in many recent camcorder UFO images. The object seen is just a light, but a shape appears on film due to distortion by the optics. Newspapers frequently publish accounts of strange-shaped camera images that are, in truth, no such thing.

22. A good case of autokinesis at work in Lincolnshire is reported in Chapter 1 of *Something in the Air*. Randles, J. Robert Hale, London, 1998.

23. *Strange But True?* Op. cit.

24. Ibid.

25. Work on the case by James Easton is published in several special editions of his on-line magazine *Voyager* at voyager@ukon-line.co.uk.

26. *Space Time Transients*. Persinger, Dr M. and Lafreniere, G. Prentice-Hall, New York, 1977.

FURTHER INFORMATION

The IUN

Much of the material in this book originated via the investigations of the Independent UFO Network (IUN). The IUN was formed in 1987 as a loose network of UFOlogists who believed that all UFO cases were explicable, and who wished to break away from the petty rules and regulations UFO investigation groups impose on their members.

The IUN investigate cases and publish their results in books, magazines and via their own magazine, *UFO Brigantia*, and the IUN website. These are available to the public. Membership of the IUN is by invitation only. IUN contacts are as follows:

Andy Roberts: brigantia@compuserve.com

Dave Clarke: crazydiamonds@compuserve.com

IUN web site: www.iun.org

Northern UFO News

Northern UFO News is the publication of NUFON, the Northern UFO Network, a network of independent UFO groups across the north and Midlands of Britain. Jenny Randles has edited the magazine since 1974. It regularly publishes the latest news, digests of cases and probable explanations, book reviews, and forthcoming lectures and conferences as well as commentary on the global UFO scene.

Northern UFO News subscriptions are available to the public at a rate unchanged in 10 years! Back issues are published freely via the IUN website. *Northern UFO News* can be contacted via Jenny Randles on nufon@currantbun.com

UFO Investigators Network

As a result of work on this book, the three authors, the IUN, NUFON and several other serious UFOlogists in the UK decided to create a new alliance for the 21st century. UFOIN was the result. This is a team of experienced investigators working to rational principles alongside scientists and sceptics as advisers. UFOIN will study cases that offer potential to add to our knowledge and re-investigate old cases seeking possible interpretations. UFOIN has decided not to operate as a UFO group and will have no members, officers, committees, nor produce magazines or hold conferences. It will solely develop a skilled, professionalised investigation team and publish in-depth case histories and research reports. It will also train novice investigators into objective ways and operate a code of practice and a complete ban on the use of regression hypnosis. The UFOIN team hopes to publish several major studies starting in the year 2000. For further details access to the web site, send an e-mail or write to UFOIN at the Buxton address below:

www.ufoin.org.uk.
enquiries@ufoin.org.uk.

Readers wishing to report a sighting or to comment on the contents of this book can contact any of the e-mail addresses or write to the authors and UFOIN c/o 1 Hallsteads Close, Dove Holes, Buxton, Derbyshire, SK17 8BS.

INDEX